Assessing [illegible] Strategies for the Joint Strike Fighter

Opportunities and Options

John Birkler, John C. Graser, Mark V. Arena, Cynthia R. Cook, Gordon Lee, Mark Lorell, Giles Smith, Fred Timson, Obaid Younossi, Jon G. Grossman

Prepared for the Office of the Secretary of Defense
Approved for public release; distribution unlimited

National Defense Research Institute RAND

The research described in this report was prepared for the Under Secretary of Defense (AT&L), within the Acquisition and Technology Policy Center of RAND's National Defense Research Institute, a federally funded research and development center supported by the Office of the Secretary of Defense, the Joint Staff, the Unified Commands, and the defense agencies under Contract DASW01-01-C-0004.

Library of Congress Cataloging-in-Publication Data

Assessing competitive strategies for the joint strike fighter : opportunities and options / John Birkler ... [et al.].
p. cm.
Includes bibliographical references.
"MR-1362."
ISBN 0-8330-3009-4
1. Joint strike fighter. 2. United States—Armed Forces—Procurement. I. Birkler, J. L., 1944– II. Rand Corporation.

UG1242.F5 A73 2001
358.4'383'072073—dc21

2001019914

Left cover photo courtesy of LockheedMartin. Reprinted by permission.
Right cover photo courtesy of The Boeing Company (www.boeing.com/jsf).
Reprinted by permission

Cover designed by Maritta Tapanainen

Published 2001 by RAND
1700 Main Street, P.O. Box 2138, Santa Monica, CA 90407-2138
1200 South Hayes Street, Arlington, VA 22202-5050
201 North Craig Street, Suite 102, Pittsburgh, PA 15213-1516
RAND URL: http://www.rand.org/
To order RAND documents or to obtain additional information, contact Distribution Services: Telephone: (310) 451-7002;
Fax: (310) 451-6915; Internet: order@rand.org

Preface

Defense policymakers in the United States expect that the Joint Strike Fighter (JSF) will play a critical role in U.S. and allied military forces through the first half of this century. As then–U.S. Defense Secretary William Cohen stated in a June 2000 letter to Congress,

> The Joint Strike Fighter (JSF) program is critically important to the modernization of United States conventional forces and is the cornerstone of tactical aircraft modernization. It will replace 1,763 Air Force, 480 Navy, and 609 Marine Corps aircraft; 2,852 aircraft in total. The JSF's stealth, advanced avionics, and ability to carry a full array of modern precision munitions will make it much more capable than the legacy aircraft it replaces when operating under challenging circumstances against modern air defenses. It is also critical to the modernization of our ally forces for coalition welfare.[1]

The Department of Defense's current JSF acquisition strategy is a "winner-take-all" competition pitting Lockheed Martin against Boeing. This strategy has raised concerns of whether competition should be retained after Lockheed Martin or Boeing is selected to begin engineering and manufacturing development.

In late July 2000, the Under Secretary of Defense for Acquisition, Technology and Logistics asked RAND to explore and identify opportunities and options to introduce competition during the production phase of the JSF. RAND researchers used as our principal criterion the likelihood that such competition would reduce the overall cost of JSF production, which is expected to total ~$300 billion in then-year dollars. We made a more limited analysis of other plausible consequences, including savings in operations and support costs, reduction in cost growth, and improvement in product quality, to the extent possible within the study's four-month duration. We also explored how new

[1] Cohen, W. S., then Secretary of Defense, letter to the Honorable Jerry Lewis, Chairman, Subcommittee on Defense, Committee on Appropriations, U.S. House of Representatives, June 22, 2000.

competitive strategies might affect foreign participation in the JSF program. U.S. allies could account for sales of an additional 3,000 JSF aircraft.

This report should be of special interest to the Office of the Secretary of Defense (OSD), service and defense agency managers, and policymakers involved in weapon system acquisitions. It was prepared for the Under Secretary of Defense (AT&L), within the Acquisition and Technology Policy Center of RAND's National Defense Research Institute (NDRI). NDRI is a federally funded research and development center sponsored by OSD, the Joint Staff, the unified commands, and the defense agencies.

Contents

Figures

Tables

Summary

The Department of Defense should stick with its winner-take-all strategy to develop and produce the Joint Strike Fighter (JSF)—which is slated to become the workhorse fighter for the Air Force, Navy, and Marine Corps. But as a hedge to ensure that later variants of this next-generation aircraft have the most effective and innovative mission systems, the Department of Defense (DoD) should consider spending money to keep a second developer and producer of these vital electronics components in the market. Such an investment in a second developer of mission systems—the electronic eyes and ears of the JSF—could be relatively modest. But it would provide future decisionmakers with the option to competitively develop a second-generation mission system source when and if it were needed.

So concludes this study of competitive strategies for the Joint Strike Fighter, performed over the summer and early fall of 2000. Constrained only by the available time (about four months), we examined both near-term and long-term competition options, some in more detail and more quantitatively than others. Throughout the study, we obtained critical information from the competing program contractors and the Joint Strike Fighter Program Office. While much of that information was considered competition-sensitive by the firms, in this report we present only our own analysis and conclusions, thus permitting unrestricted distribution.

The bottom line is that putting one company in charge of the overall production of the JSF makes the most economic sense. If two or more competitors develop and build this next-generation aircraft, the DoD likely would not see lower overall program costs. The high front-end investments, together with increases in recurring costs, needed for producing such a sophisticated weapon system probably would not be recovered through price reductions that might result from competitive forces.

At the same time, however, having several companies funded to develop and potentially produce later variants of the JSF mission system also may make sense. With technology changing at an increasing pace, the DoD could face the prospect of having its next-generation fighter employ less than the most innovative technologies. By funding competitive technology development at the mission system level, defense policymakers might be able to lessen that risk.

The Problem

The JSF program is a joint effort among the U.S. Air Force, U.S. Navy, and U.S. Marine Corps, with full partnership participation by the United Kingdom (UK). The program objective is to develop and deploy a family of strike aircraft to meet the operational needs of the Air Force, Navy, Marine Corps, United Kingdom, and other U.S. allies. The different variants are expected to have a high degree of commonality, thus reducing overall acquisition and support costs.

The JSF will be one of the largest acquisition projects in history, worth some $300 billion (then-year dollars), and the only new major fighter aircraft program planned for the next 30 years. Over the next several decades, these aircraft are slated to replace all F-16s, A-10s, and AV-8Bs in the U.S. inventory, the F/A-18C/Ds in the Marine Corps inventory, and Harriers in the UK inventory; and to augment the Navy's F/A-18E/Fs. By 2026, some 3,000 of the jet fighters are planned to be integrated into U.S. and UK forces. The DoD expects that additional sales to U.S. allies could approach 3,000 aircraft.

In fall 2001, the DoD intends to grant one prime contractor the right to develop and produce all variants of the JSF. In 1996, the DoD named the Boeing Company and the Lockheed Martin Corporation as the JSF's two finalist prime contractor competitors. Since then, each has been engaged in concept demonstration efforts, flown test vehicles, and pursued other extensive preliminary development efforts—all in hopes of winning the JSF contract.

The DoD has been using this "winner-take-all" approach for decades; it has been the preferred method of developing and producing sophisticated weapon systems. But as Boeing's and Lockheed Martin's efforts have unfolded, senior DoD officials and members of Congress have voiced concerns that awarding the JSF to one company could lead to a situation in which the United States might be paying more for weapon systems or purchasing less technologically sophisticated platforms than it would under more robust competition.

What RAND Was Asked to Do About the Problem

In late July 2000, the Under Secretary of Defense for Acquisition, Technology and Logistics asked RAND to explore and identify opportunities and options to introduce competition during the production phase of the JSF. RAND researchers used as our principal criterion the likelihood that such competition would reduce the overall cost of JSF production. We made a more limited analysis of other plausible consequences, including savings in operations and support costs, reduction in cost growth, and improvement in product quality, to the extent possible within the study duration of approximately four months. We also explored how new competitive strategies might affect foreign participation in the JSF program.

How RAND Studied the Problem

The most direct approach to estimating the results of introducing competition would be to estimate the cost for a sole-source producer, then estimate the cost for a pair of competitive producers, and compare the two. Unfortunately, we have no data or cost-estimating relationships that would enable us to directly estimate production costs in a competitive environment. What we do have is some historical data showing the amount by which production costs *changed* when competition was introduced into ongoing sole-source production programs. Therefore, in this study we estimate the likelihood that the government would "break even" on the introduction of a competitive second source; that is, that the cost reductions would be great enough to pay for the incremental costs of introducing competition. If the likelihood is high, the government might reasonably elect to introduce competition in the expectation of achieving other potential benefits. Likewise, a low expectation of breaking even on production cost would discourage the government from introducing competition, because the net dollar cost of production might increase enough to outweigh other possible benefits.

In our quantitative analysis, we developed a "break-even" model, which was based on previous RAND studies that examined how competition might be introduced into the development and production phases of a variety of weapon systems. Our model was developed specifically for the JSF program and takes into consideration the multiple, unique aspects of the program. Using this model, we were able to gauge the likelihood that the DoD would recoup its costs if it invested in a second JSF producer. In particular, the model allowed us to look at whether lower JSF production costs engendered by the presence of a second producer would offset DoD's investment in bringing that second source into the picture. This quantitative analysis relied on three main sources of information:

- Proprietary estimates from Boeing and Lockheed Martin of their designs' development and production costs, at a level of detail that enabled us to estimate costs under different production scenarios.
- JSF Program Office cost estimates for each variant, together with overall programmatic information on past and projected schedules of events, production quantities, and other information.
- RAND data on cost and cost-estimating relationships for development, production, and operation of U.S. fighter aircraft.

We augmented these sets of quantitative data with extensive discussions with both contractor teams and the project office staff regarding the feasibility and desirability of various competition strategies. This qualitative analysis allowed us to gauge whether the prospect of saving overall costs via a particular competition path was reasonable. We used the following to flesh out this part of our analysis:

- Studies and analyses on the results of prior efforts to introduce competition to a weapon production program. The time available for this study was limited. Therefore, our analysis of the historical record was limited to a review of secondary sources—

reports on previous studies of production-cost changes resulting from competition. We examined the DoD's experience with introducing competition in 62 weapon system and component procurement programs between 1960 and 1990.

While useful, those reports rarely provided all of the information we needed to apply the historical results to the particular conditions presented by the JSF program. Nevertheless, we were able to assemble a body of historical data sufficient to support conclusions with useful confidence and precision.

Our Findings and Conclusions

Whether and how much a second producer of a weapon system generates cost savings for the DoD depends on the type of hardware or system that the competitors are developing and manufacturing. The potential for cost savings also depends on the time frame—near term or far term—in which the competition is taking place.

Our examination of the DoD's past experience with introducing competition into weapon system procurements suggests that second producers of electronics have been more likely to generate savings in production costs than have second producers of missiles and ships. As Table S.1 shows, half of the DoD programs in our historical survey that involved two or more competitive producers of electronics were able to reduce overall costs by 30 percent, but only one in 10 competitive missile and ship production efforts was able to do so.

Table S.1
Fraction of Competitive Programs Examined That Achieved Savings

Savings achieved (percent)	Missiles and ships	Electronics
>0	7/10	9/10
>10	5/10	8/10
>20	3/10	6/10
>30	1/10	5/10
>40	Nil	3/10

RAND*MR1362-TS.1*

But such a 30-percent cost reduction is just the level of savings the DoD would need to achieve if it were to bring a second competitor into the JSF program. Our break-even analysis suggests that such a second competitor would need to offer the DoD this level of savings across a range of the JSF's elements—forward, center, and aft fuselages; landing gear; ejection seat; wing; edges; empennage; and mission systems. Given DoD's past experience with missiles and ships, we found that such savings from competition would be difficult, if not impossible, for the JSF program to achieve in the near term in any of these areas.

With the exception of the mission system, all these areas would be equally unpromising in the far term (a decade or more in the future). The mission system constitutes the eyes and ears of the JSF and provides a powerful contribution to its overall combat effectiveness. Many of the enabling technologies are evolving much more rapidly than most flight-vehicle technologies, and several major upgrades in mission systems are likely during the life of the JSF. It therefore makes sense to begin preparing now for the first upgrade. We suggest that the DoD investigate establishing a "shadow" industry team to begin developing system architectures and component technologies that would be tailored to the JSF but focused on technological advancement, cost reduction, and any new mission requirements that might emerge. That team would then be capable of competing to develop and produce an integrated suite of mission equipment for a future upgraded JSF.

Policy Implications That Derive from Our Findings

Our findings have two policy implications:

- *Policymakers should stick with the winner-take-all strategy for near-term development and production of the JSF.* Despite the potential advantages that might accrue, establishing a competitive production line for part or all of the JSF would require a front-end investment, together with increases in recurring costs, that probably would not be recovered through price reductions that might result from competitive forces.

- *Policymakers should examine the costs and benefits of supporting a second industry team so that it could be capable of competing to develop and manufacture the next major upgrade of the mission system equipment.* If implemented, this strategy would ensure that future managers have the option of a competitive second source, one that might not otherwise be available.

Our charter was to explore opportunities and options to inject competition into the JSF production program, with the principal criterion being the likelihood that such competition would reduce the overall cost of JSF production. Our analysis shows that introducing competition for future upgrades to the mission system presents the most attractive option. Within the scope of our study, we were unable to examine the idea in detail; we therefore recommend only that the notion be examined with an appropriate level of care and detail by the JSF management.

Acknowledgments

This work could not have been undertaken without the special relationship that exists between the Office of the Secretary of Defense (OSD) and RAND under the National Defense Research Institute (NDRI). For that relationship we are grateful. Many individuals in OSD, British Ministry of Defence (MOD), Joint Strike Fighter System Program Office, and RAND deserve credit for the work discussed in this report. Their names and contributions would fill several pages.

If we were to single out a senior person at OSD, MOD, JSF Program Office, and RAND who participated in and supported this work in extraordinary ways, we would mention Dr. Jacques S. Gansler, Under Secretary of Defense (Acquisition, Technology and Logistics); Sir Robert Walmsley, Chief of Defence Procurement and Chief Executive, Defence Procurement Agency; Col.(s) Darrell H. Holcomb, USAF, Director of Acquisition Strategy, Joint Strike Fighter Program; and Dr. Robert Roll, Program Director, Resource Management Program at RAND.

We also want to thank the leadership and staff of the Office of the Secretary of Defense, the British Ministry of Defence, the Joint Strike Fighter Program Office, and Boeing, Lockheed Martin, Northrop Grumman, Raytheon, BAE Systems, Rolls-Royce, Smiths Industries, and Flight Refuelling. Each gave us the opportunity to discuss a broad range of issues with those directly involved; in addition, several of the firms arranged for us to visit their facilities. The firms and government offices provided all the data we requested in a timely manner.

This broad-based participation made possible the analysis described here.

We wish to thank RAND colleagues Frank Lacroix and Donald Stevens. Their thoughtful reviews occasioned many changes that improved the clarity of the report.

Lastly, the authors owe RAND colleague Joan Myers an incalculable debt for her thorough and patient administrative assistance at every stage in the project.

Abbreviations and Acronyms

2D	Two-dimensional
3D	Three-dimensional
ACM	Advanced Cruise Missile
AESA	Active electronically scanned array
AEP	Alternate engine program
AT&L	Acquisition, Technology and Logistics (Under Secretary of Defense)
BTP	Build-to-Print
CAIV	Cost As an Independent Variable
CDP	Concept Demonstration Phase
CGF	Cost-growth factor
CTOL	Conventional takeoff and landing
CV	Carrier variant
CY	Calendar year
DCASPRO	Defense Contract Administration Services Plant Representative Office
DoD	Department of Defense
EMD	Engineering and Manufacturing Development
ESSS	Electronic Sensors and Systems Sector (Northrop Grumman)
EW	Electronic warfare
FACO	Final assembly and check-out
FFF	Form-Fit-Function
FLIR	Forward-looking infrared

FMS	Foreign Military Sales
FY	Fiscal year
GD/C	General Dynamics Convair
HPCS	Historical production-cost savings
ICP	Integrated core processor
IOC	Initial operational capability
IPT	Integrated Product Team
JASSM	Joint Air-to-Surface Standoff Missile
JIRD	Joint initial requirements document
JSF	Joint Strike Fighter
LO	Low observable/observability
LOA	Letter of Offer and Acceptance
LRIP	Low-rate initial production
MIRFS	Multifunction integrated radio frequency systems
MoA	Memorandum of Agreement
MOD	Ministry of Defence (United Kingdom)
MoU	Memorandum of Understanding
NAVAIR	Naval Air Systems Command
NDRI	National Defense Research Institute
ORD	Operational Requirements Document
O&S	Operations and support
OSD	Office of the Secretary of Defense
PA&E	Program Analysis and Evaluation
PDRR	Program Definition and Risk Reduction
PHM	Prognostic Health Monitoring
RAF	Royal Air Force (United Kingdom)
RCR	Required cost reduction
RFP	Request for proposal
SINCGARS	Single-Channel Ground and Airborne Radio System

STOVL	Short takeoff/vertical landing
TDP	Technical data package
UCAV	Unmanned Combat Air Vehicle
UK	United Kingdom
URF	Unit recurring flyaway
USAF	United States Air Force
USMC	United States Marine Corps
USN	United States Navy

Chapter 1
Introduction

Between July and October 2000, RAND researchers analyzed options open to the Department of Defense to inject competition into various stages of production of the Joint Strike Fighter (JSF), the only new major fighter program planned for the next 30 years. Done quickly at the request of the Under Secretary of Defense for Acquisition, Technology and Logistics, this study examined ways that defense policymakers may be able to lower the aircraft's costs or improve its quality by introducing competition during production.

Introducing competition into the JSF production process will not be easily or quickly done. The DoD intends in fall 2001 to grant one prime contractor the right to develop and produce all variants of the new fighter—a production run that could exceed 3,000 aircraft, extend to 2026 or beyond, and be worth $300 billion or more in then-year dollars. The DoD in 1996 selected Boeing and Lockheed Martin as the JSF's two finalist prime contractor competitors. Since then, each has been engaged in concept demonstration efforts, flown test vehicles, and pursued other extensive preliminary development efforts—all in hopes of winning the JSF contract.

For decades, this winner-take-all approach has been the DoD's preferred method of developing and producing sophisticated weapon systems. But as Boeing's and Lockheed Martin's efforts have unfolded, senior DoD officials and members of Congress have voiced concerns that awarding the JSF to one company could lead to a situation in which the United States might be paying more for weapon systems or purchasing less technologically sophisticated platforms than it would under more robust competition.

The main focus of the study RAND was asked to undertake in July 2000 was estimating the likelihood that introducing competition would reduce the overall cost of JSF production. To the extent possible within the study duration of approximately four months, we made a more limited analysis of other plausible consequences, including savings in operations and support costs, reduction in cost growth, and improvement in product quality. We also explored how new competitive strategies might affect foreign participation in the JSF program.

Background

The JSF will be one of the largest acquisitions in history, worth some $300 billion (then-year dollars) over the next quarter century. Some 3,000 aircraft—configured in a conventional takeoff and landing (CTOL) variant, a short takeoff and vertical landing (STOVL) variant, or a carrier variant (CV)—will be produced for the DoD. The JSF Program Office estimates that foreign customers could purchase an additional 3,000 aircraft.

The JSF was conceived in the early 1990s as an aircraft that would meet the long-term needs of the three major U.S. services that operate fighter aircraft by

- replacing U.S. Air Force F-16s and A-10s
- augmenting carrier-based U.S. Navy F/A-18E/Fs
- replacing U.S. Marine Corps F/A-18C/Ds
- replacing U.S. Marine Corps AV-8Bs and United Kingdom (UK) Harrier aircraft.

The JSF will be the first fighter program that attempts to satisfy the needs of three services with one basic design of high commonality among multiple variants. The goal is to have some 70-percent commonality in structure and subsystems, a common core engine, and very high commonality in mission systems (avionics) across all three variants. In theory, this approach should make the JSF more affordable during production and throughout the service life of the aircraft. The DoD has established target flyaway costs for each variant: $28 million for the CTOL, $30–$35 million for the STOVL, and $31–$38 million for the CV (in 1994 $). Some strategies for achieving commonality are depicted in Figures 1.1 and 1.2, the Boeing and Lockheed Martin variants of the platform.[1]

The notion of a "common" design that would satisfy the diverse operational needs of Air Force, Navy, and Marine Corps poses an exceptional challenge to the designers. Only once before in recent history has an attempt been made to design one basic aircraft with variants that would perform land-based and carrier-based operations. That attempt failed when the carrier-based variant of the F-111 was abandoned.[2]

RANDMR1362-1.1

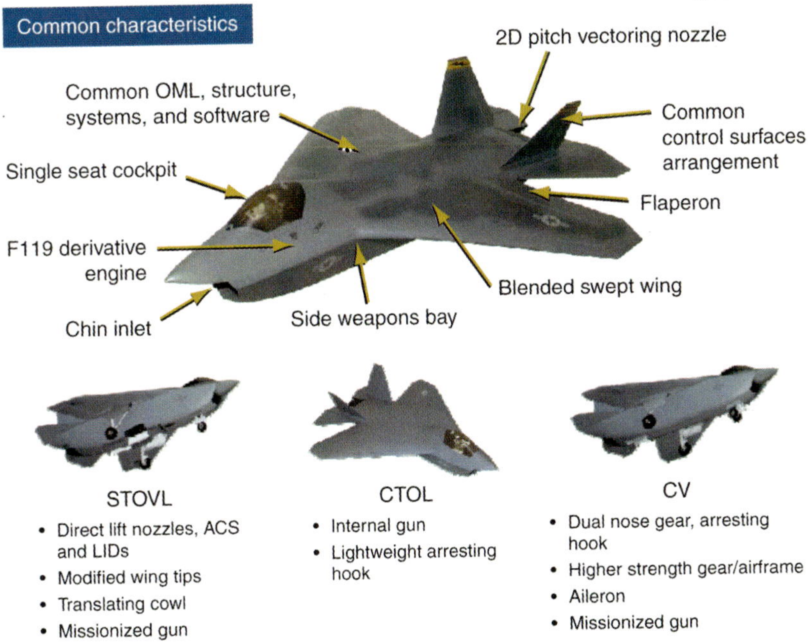

OML = outer mold line; ACS = attitude control system; LID = lift improvement device.

SOURCE: Joint Strike Fighter Program Office.

Figure 1.1—Boeing JSF Design

---

[1] The designs shown represent the pre-proposal versions of the operational JSF and do not necessarily reflect the JSF configurations that will be proposed by the two firms.

[2] Some carrier-based designs were later modified or adapted to land-based operations with great success—for example, the F-4 platform developed for carrier-based operations and subsequently adapted by the Air Force for land-based operations.

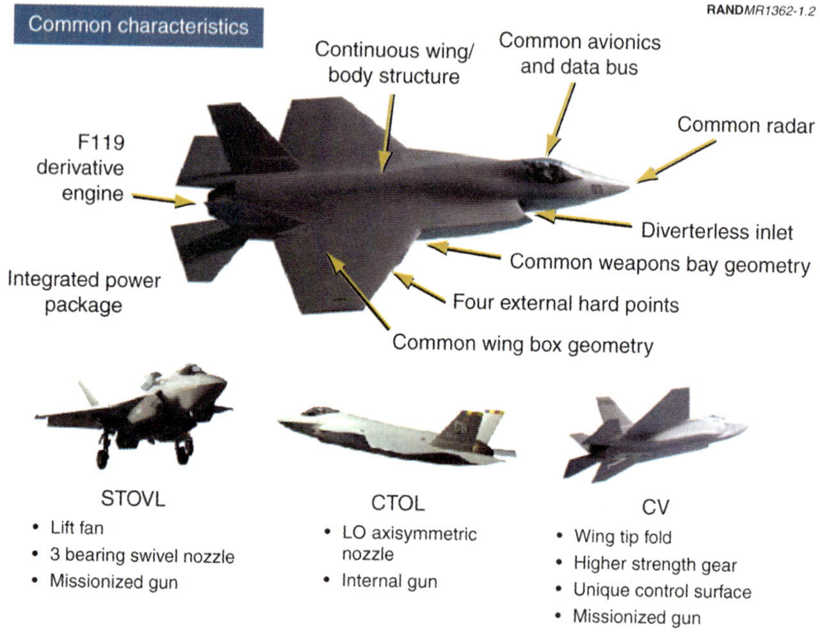

SOURCE: Joint Strike Fighter Program Office.

Figure 1.2—Lockheed Martin JSF Design

Those challenges have led to an unusually long and thorough period of requirements formulation, design analysis, and technology demonstration in the JSF program. As shown in the JSF program schedule in Figure 1.3, the contractors' concept demonstrations began in FY97 and will continue until late FY01. Both Boeing and Lockheed Martin flew test variants of their aircraft in late CY00, and more test flights are slated through mid-2001.

Once the DoD chooses the JSF prime contractor, Engineering and Manufacturing Development will begin. Delivery of the first operational aircraft is expected in FY08.

Study Objectives

RAND's involvement with this issue and its study objectives stem from a June 2000 letter that then–Secretary of Defense William Cohen wrote to Representative Jerry Lewis, Chairman of the Subcommittee on

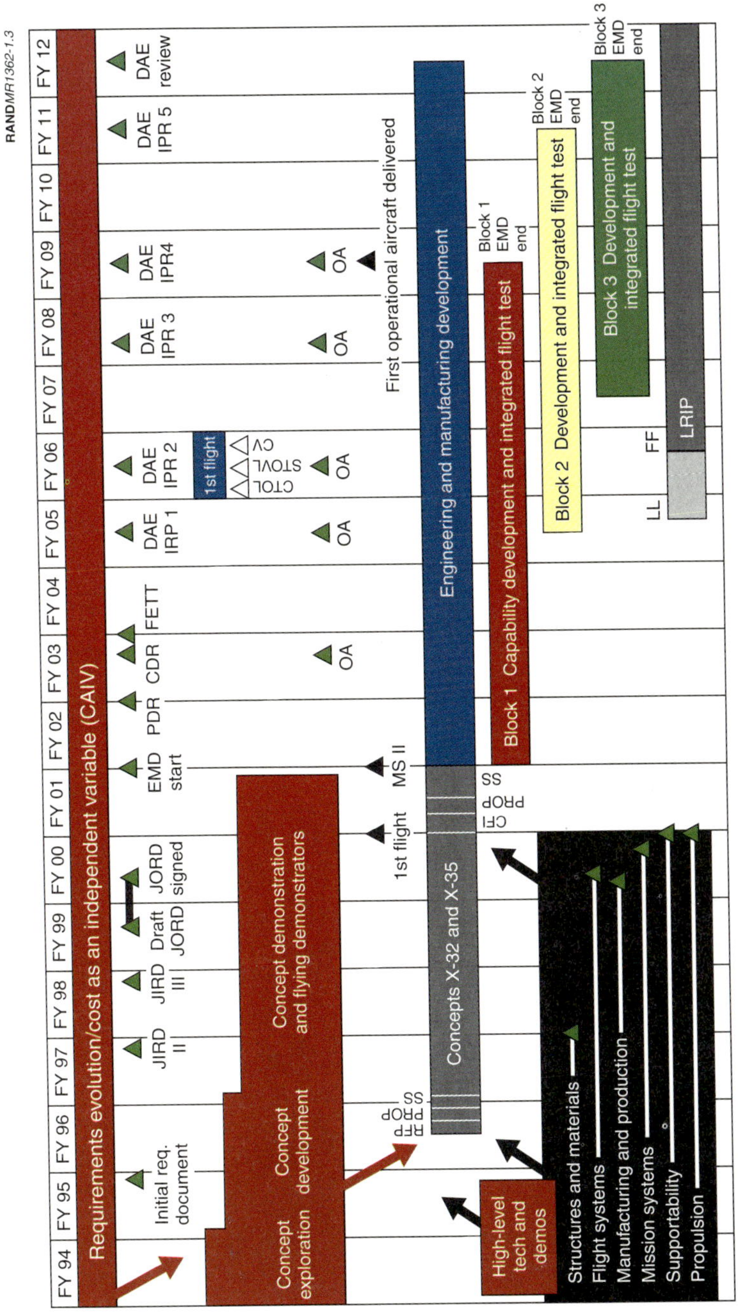

SOURCE: Joint Strike Fighter Program Office, July 2000.

Figure 1.3—Joint Strike Fighter Program Schedule

Defense in the House Appropriations Committee. "The Department has examined a number of options for continuing the JSF program once concept demonstration is completed," Secretary Cohen wrote. "These options all assume the selection of a single winning design. They range from winner-take-all to competition throughout production. We will continue to evaluate these options and to develop a comprehensive assessment that our successors can use as they make decisions on the future course of the JSF program. I have asked the RAND Corporation to review these options, including cost, and report back by December 1st."

Sources We Drew Upon to Meet the Study Objectives

Our analysis was based on four main sources of information:

- First, each of the prime contractor teams provided its own (proprietary) estimates of development and production costs for its individual designs, at a level of detail that enabled us to estimate costs under different production scenarios.
- Second, the JSF Program Office provided its own cost estimates, together with overall programmatic information on past and projected schedules of events, production quantities for each variant, etc.
- Third, we drew on our own internal stores of cost data and cost-estimating relationships for development, production, and operation of U.S. fighter aircraft, and we augmented these sets of quantitative data through extensive discussions with both contractor teams and with the Program Office staff regarding the feasibility and desirability of various competition strategies.
- Fourth, we reviewed the body of literature on the results of prior efforts to introduce competition to a weapon production program. The short time available for this study (about four months from go-ahead to final briefing) limited our review of the historical record on the effects of competition on production costs to secondary sources—reports on previous studies of production-cost changes due to competition. While useful, those reports

rarely provided all of the information we needed to apply the historical results to the particular conditions presented by the JSF program. Nonetheless, we were able to assemble a body of historical data sufficient to support conclusions of useful confidence and precision.

Limitations on the Scope of the Study

Competition is generally expected to stimulate a wide variety of actions by the producers in attempting to make their product more attractive to the buyer. This study was almost entirely focused on two possible consequences of competition: changes in production costs and changes in operations and support costs. Several other possible consequences of production are briefly reviewed. They are factored into the overall conclusions, but only the costs of production and of operations and support are examined quantitatively.

Unfortunately, we have no historical or analytical methods for directly estimating production cost in a competitive environment. Instead, we estimated the incremental costs of introducing competition; then, drawing on historical evidence, we attempted to estimate the likelihood that competition would drive down the costs enough to permit recovering those incremental costs, thus allowing the government to at least break even on costs. The short time available for the study of previous competitive production programs limited us to a review of surveys done by others; we were unable to examine the original records of those past programs. We did not evaluate the engine competition, which is already an integral part of the JSF acquisition strategy.

Report Organization

The report is organized in ten chapters. Following this Introduction, Chapter Two describes the overall JSF project. Chapter Three discusses some of the special features of competition in production of weapon systems and presents the view of such competition as a balance between benefits and liabilities. Chapter Four summarizes the overall analysis process employed in this study. Chapters Five and Six describe the results of our analysis of the effects of introducing competition on production cost and on operations and support costs, respectively. Chapter Seven briefly reviews some of the likely benefits of competition other than costs. Chapter Eight summarizes the near-term options for introducing competition in the JSF program, and Chapter Nine explores some options for introducing competition in later phases of the program and for competition from outside the program. Final conclusions and recommendations are presented in Chapter Ten.

Chapter 2

Description of the Joint Strike Fighter Program

The Joint Strike Fighter (JSF) program is expected to be one of the largest and most costly acquisitions in history. Current plans call for the armed forces of the United States and United Kingdom to procure 3,002 JSF aircraft, at a cost of more than $300 billion (then-year $). In an attempt to achieve commonality, variants of the JSF are planned eventually to replace a large percentage of the major fighter/attack aircraft in the current inventories of the U.S. Air Force, Navy, and Marines (in addition to British Royal Navy and Royal Air Force [RAF] Harriers). This replacement is illustrated in Figure 2.1.

RANDMR1362-2.1

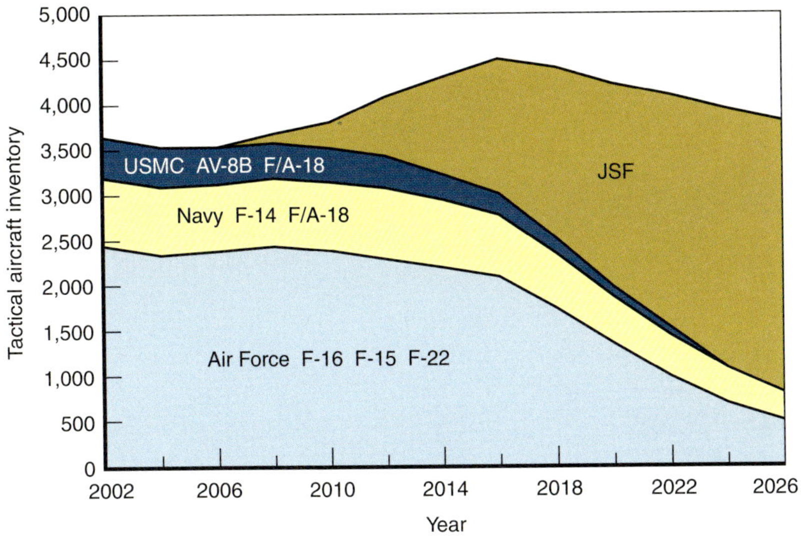

Figure 2.1—Projected Fighter/Attack Aircraft Inventory for U.S. Forces

Following the completion of the current Concept Demonstration Phase (CDP)—during which Boeing and Lockheed Martin, the two competing contractors, will fly concept demonstration aircraft (designated X-32 and X-35, respectively)—DoD plans to select a single prime contractor in a winner-take-all strategy intended to maximize commonality among variants and achieve maximum economies of scale. Therefore, the JSF program also has important implications for the industrial base. This aircraft could end up as the only major fighter aircraft in production in the United States by 2020 or earlier, as shown in Figure 2.2. This compares with the early years of the current decade, which should witness the simultaneous production of at least three fighters (F/A-18E/F, F-22, and variants of the F-16 and/or F-15). Thus, the U.S. government could find itself relying on only one credible source for designing, developing, and manufacturing fighter aircraft by the 2020s.

The JSF is particularly important for the U.S. Marine Corps. Currently, the Marine fixed-wing tactical aircraft inventory is made up of

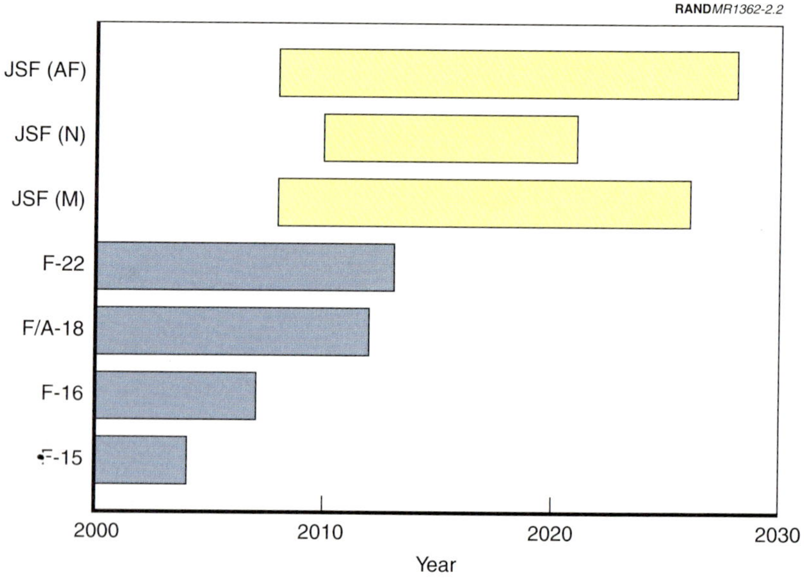

Figure 2.2—Projected Production Schedules for U.S. Tactical Aircraft

about 175 Boeing (McDonnell Douglas)/BAE Systems[1] AV-8B Harriers, a like number of Boeing (McDonnell Douglas)/Northrop Grumman F/A-18C/Ds, and less than 100 F/A-18A/Bs. All of these aircraft are expected to be retired from the inventory between 2015 and 2021. Plans call for the delivery of the first STOVL JSF variant to the Marines in 2008. By 2011, deliveries are planned to stabilize at their peak level of 36 Marine STOVL variants per year, for an eventual total of 609 aircraft. Thus, by 2021, the STOVL JSF is expected to be the only tactical fighter aircraft in the Marine inventory.

The JSF is also crucial for U.S. Air Force modernization requirements. Indeed, the Air Force will be the single-largest customer for the program. The current USAF tactical fighter/attack inventory includes about 2,500 aircraft, made up of nearly 1,400 Lockheed Martin (General Dynamics) F-16s, about 500 Boeing (McDonnell Douglas) F-15s and 200 F-15Es, about 50 Lockheed Martin F-117 stealthy attack aircraft, and about 340 Fairchild Republic A/OA-10 attack aircraft. To replace the F-15 as its premier air superiority fighter, the Air Force plans to buy 339 Lockheed Martin F-22 air superiority fighters. Beginning in 2008, the Air Force expects to begin receiving 1,763 conventional takeoff and landing (CTOL) variants of the JSF to replace the F-16s and A-10s. By the mid-2020s, the JSF will make up about 70 percent of the Air Force tactical fighter inventory.

The JSF also plays a key role in the Navy's aviation modernization plans. The current Navy inventory is made up of fewer than 250 aging Northrop Grumman F-14 fleet air defense fighters, about 150 Boeing (McDonnell Douglas) F/A-18A/B fighter/attack aircraft, and about 350 F/A-18C/D fighter/attack aircraft. Beginning in 1998, the Navy began procuring the first of a planned 548 F/A-18E/Fs to replace the F-14s, which will be phased out of the inventory by 2008, and the F/A-18A/Bs, which will be gone by 2015. To maintain the full force-structure size requirement as the F/A-18C/Ds begin retiring in large numbers by 2010, the Navy plans to start procurement of the carrier variant (CV) JSF beginning that same year. By the mid-2020s, the CV

[1] British Aerospace became BAE Systems on November 30, 1999.

JSF is planned to make up approximately 60 percent of the Navy tactical fighter force structure. Full procurement of the planned 480 CV JSFs is necessary to meet the Navy tactical fighter force-structure requirement.

Finally, STOVL variants of the JSF could replace 150 RAF GR.7 Harriers and Royal Navy Sea Harriers. The JSF and the Eurofighter are expected to become the principal tactical air assets of the UK.

Original Acquisition Plan and Program Priorities

From the early phases of the JSF program, the intent of acquisition planners was to structure the program differently from past programs in order to reduce overall costs of ownership. DoD designated JSF a flagship program for acquisition reform. Changes to traditional acquisition approaches concentrated on six broad areas: service commonality, the acquisition cycle, the requirements determination process, technical risk reduction, extended design and subsystem competition, and foreign participation.

Service Commonality

Early in the program it was decided that, to reduce costs through increasing economies of scale and to promote interoperability, variants of the same basic aircraft design would be used to satisfy the requirements for the tactical fighter modernization of three services: Navy, Air Force, and Marine Corps. The competing contractors were encouraged to maximize commonality among their three variants so that they could use the maximum possible number of identical or similar airframe parts, as well as avionics, engine, and other subsystems. The intent was that all three variants could be assembled using a large number of common or similar parts and components.

Acquisition Cycle

Planners established a Program Office much earlier than is usual, with numerous Integrated Product Teams (IPTs) representing most of the major stakeholders from the acquisition, user, and support communities. In addition, the traditional Concept Exploration and Program Definition and Risk Reduction (PDRR) phases were combined into a single Concept Demonstration Phase (CDP).[2] The longer CDP permitted a much lengthier period to conduct more-extensive cost/benefit trade-off analyses, technology assessment, and requirements definition.

Requirements Determination

The requirements determination process was far more drawn out and iterative than in a traditional DoD acquisition program. From FY95 through FY99, three joint initial requirements documents (JIRDs) were developed through an iterative process that included all stakeholders. Promoted during this period was vigorous application of the acquisition reform concept of Cost As an Independent Variable (CAIV), which raises cost goals to the same level as performance and other system requirements. Indeed, specific Unit Recurring Flyaway (URF) cost threshold targets were established early in the program and became key requirements: $28 million for the CTOL variant, $31–$38 million for the carrier variant, and $30–$35 million for the STOVL variant (costs in FY94 $). To accommodate these targets, the user communities had to make trade-off decisions on the costs and benefits of various performance requirements. The process of developing initial requirements documents led in FY99 to the writing of a draft Operational Requirements Document (ORD) and a subsequent iteration, resulting in a final ORD in mid-FY00.

The two U.S. lead prime contractors and their major partners have an unprecedented degree of design and configuration control and responsibility. The ORD states primarily broad performance parameters and URF costs. The contractors have been granted considerable freedom in developing designs, selecting subcontractors, and choosing technologies to meet the performance and cost requirements. For example, each prime contractor has selected different technological approaches to the STOVL lift system, as well as to many other sys-

[2] To be completely accurate, the Joint Advanced Strike Technology (JAST) program, which preceded the JSF program, did have a Concept Exploration phase, which lasted through most of CY94. Rather than being formally viewed as an acquisition program, JAST was viewed as a technology exploration program. The main achievement of the JAST Concept Exploration phase was the conclusion that variants of a single basic design could meet the tactical fighter modernization requirements of all three U.S. services. In addition, the first part of the CDP beginning in November 1994 was originally referred to as the Concept Definition and Design Research Phase.

tems, such as the electro-optical/infrared systems, the fire control radar, and mission computer architecture. Thus, the contractors are responsible for achieving the required capabilities (performance, reliability, supportability, etc.) within the cost goals established by the government, but the way they do so is left up to them.

Risk Reduction

Early on, the JSF Program Office identified areas of relatively high technical and programmatic risk, and initiated programs to reduce risk. The main approach selected to manage risk was to fund numerous competitive hardware demonstration programs. One good example is the multifunction integrated radio frequency systems (MIRFS) program.

The purpose of the MIRFS program was to encourage companies to develop lighter, much lower cost, active electronically scanned arrays (AESAs) for fire control radars. In an unprecedented risk-reduction effort, the JSF Program Office funded a five-year, $110-million program beginning in February 1996. Contracts were let to the two leading U.S. fire control radar contractors, Hughes Aircraft, which later merged into Raytheon, and Northrop Grumman Electronic Sensors and Systems Sector (ESSS). During CDP, Raytheon was the radar source to Boeing and Northrop Grumman/ESSS was the supplier to Lockheed Martin. Both companies undertook major efforts to develop new technology and manufacturing processes to dramatically reduce the cost and weight of AESAs. The programs included flight demonstrations of prototype arrays incorporating the new technologies.

The JSF Program Office funded many other, similar technology risk-reduction programs, including the structures and materials programs, the avionics virtual systems engineering prototyping program, and the joint visual system operational evaluation program.

These technology-management and risk-reduction programs differed in several ways from past programs. First, they were more directly linked to specific preferred weapon system concepts being developed by the prime contractors. Second, they often included two competing contractors undertaking the same type of R&D. Third, they were often larger, longer, and more generously funded than similar types of programs in the past. Fourth, they often required development and a full-scale flight demonstration of the actual hardware, rather than laboratory demonstrations.

Of course, the most dramatic example of this approach to risk reduction is the requirement of the JSF's two competing prime contractors to develop and flight-test demonstration aircraft, as discussed below.

Competition

Throughout the CDP, competition at the prime and subsystem levels was emphasized. During the first year of the CDP,[3] three prime contractor teams competed: Boeing, Lockheed Martin, and McDonnell Douglas/Northrop Grumman/British Aerospace.[4] In November 1996, DoD selcted Boeing and Lockheed Martin to complete the CDP. These two prime contractors would continue to compete until the selection of a single winner-take-all for the Engineering and Manufacturing Development (EMD) phase, now expected to begin in fall 2001. In late 1995, the Alternate Engine Program (AEP) was launched with a $7-million contract awarded to General Electric teamed with Allison (Rolls-Royce). Originally, the Pratt & Whitney F119 engine had been selected for use by all of the prime contractors for their JSF designs. The AEP provided money to complete the full-scale development of the GE F120, so that it would be available as a competitive alternative to the P&W F119 after the initial JSF production phases are completed in the early 2010s. In a like manner, competitive technology programs were funded for many other major subsystems, such as the fire control radar.

Each contractor is building and flight testing two demonstrators. Both Boeing's X-32 and Lockheed's X-35 first flew in the second half of 2000. These aircraft are intended to demonstrate basic design charac-

---

[3] Also known as the Concept Definition and Design Research Phase, which lasted from November 1994 to November 1995.

[4] After the start of CDP, Northrop Grumman and British Aerospace (British Aerospace became BAE Systems on November 30, 1999) joined the Lockheed Martin team, and Boeing merged with McDonnell Douglas.

teristics in three areas: commonality and modularity for cost-effective variants for the three services, short takeoff and vertical landing, and low-speed carrier approach and flying capabilities.

International Participants

Unlike most other major DoD acquisition programs, JSF has included significant foreign participation from the very beginning of the design phase—something no U.S. fighter developed since World War II has included. Foreign government and industry participants have been included for the following reasons: to enhance equipment interoperability with allies, to promote foreign acquisition of the aircraft, to share the financial burdens of development and production, to share U.S. know-how with important foreign allies, and to gain access to unique technologies and capabilities available from key allies.

There have been four official levels of foreign government participation during the CDP:

- Full Collaborative Partner
- Associate Partner
- Informed Partner
- Foreign Military Sales (FMS) Major Participant.

Full Collaborative Partner. The United Kingdom is the foreign participant with the most significant involvement in the program. It is the only Full Collaborative Partner on the program for the CDP portion. The U.S. and UK governments signed a Memorandum of Understanding (MoU) on Royal Navy program participation in 1995. The MoU was modified in 1999 to include the Royal Air Force. The UK is contributing $200 million to the CDP.

The UK actively participates in the process of developing JSF requirements documents. The stated objective of the U.S.–UK MoU is to promote harmonization between UK system requirements and the requirements of the U.S. services. UK personnel are included on several IPTs. By the later stages of the CDP, the UK held eight country representative positions at the Program Office, in addition to a National Deputy at the Director level.

Associate Partner. Three countries have Associate Partner status during CDP: Denmark, the Netherlands, and Norway. These three countries jointly negotiated agreements that were signed on different dates in 1997. Each country contributed $10 million, which was matched by a U.S. contribution of $30 million, for a total of $60 million. The primary objective of their participation in the CDP is to influence requirements development for the CTOL JSF variant. In line with the terms of the MoUs and Memoranda of Agreement (MoAs), Associate Partners may influence requirements as long as they and the United States perceive the results to be mutually beneficial. Each Associate Partner is represented by one National Deputy and one technical representative during the CDP.

Informed Partner. Canada and Italy participate as Informed Partners. The United States signed an MoU with Canada in January 1998, and an MoA with Italy in December of the same year. Informed Partners do not have the authority to influence requirements. Canada participates in design refinements of the CTOL variant, and other associated activities. Italy is involved in a variety of tasks related to the Italian Navy's interest in the STOVL variant and the Italian Air Force's interest in the CTOL variant. Canada and Italy are each contributing $10 million toward this joint activity. The United States contributes $50 million to the joint U.S.–Canadian activities.

Foreign Military Sales Major Participant. There are three FMS Major Participants in the CDP: Turkey, Singapore, and Israel. All three signed Letters of Offer and Acceptance (LOAs) during 1999. All FMS Major Participants are involved in the generic JSF project, which provides extensive unclassified and nonproprietary information about JSF requirements and designs. Each of the three FMS Major Participants takes part in a variety of different aspects of the CDP. Turkey committed to a contribution of $6.2 million; Singapore and Israel are contributing $3.6 million and $0.75 million, respectively. There is no U.S. financial contribution to these joint FMS Major Participant efforts. A summary of CDP foreign participation is shown in Table 2.1.

As of late 2000, negotiations were under way with many of these countries, as well as others, regarding the possibility of developing agreements for participation in EMD. According to press accounts,

Table 2.1
International Participation in JSF CDP

Country	Status	Agreement	Foreign contributions	U.S. contributions	Date joined
United Kingdom	Full Partner	MoU	$200M	—	Dec 95
Netherlands	Associate Partner	MoA	$10M	$10M	Apr 97
Norway	Associate Partner	MoU	$10M	$10M	Apr 97
Denmark	Associate Partner	MoU	$10M	$10M	Sept 97
Canada	Informed Partner	MoU	$10M	$50M	Jan 98
Italy	Informed Partner	MoA	$10M	—	Dec 98
Singapore	Major Participant	LOA	$3.6M	—	Mar 99
Turkey	Major Participant	LOA	$6.2M	—	Jun 99
Israel	Major Participant	LOA	$0.75M	—	Sep 99

RAND*MR1362-T2.1*

four levels of EMD participation will be available for foreign partners: Levels 1, 2, and 3, and FMS. Level 1 will require approximately a 10-percent contribution to EMD costs. Level 2 will require about 5 percent, and Level 3 will require on the order of 1 to 2 percent.

The United Kingdom will participate in the evaluation of the CDP flight demonstrations, as well as in the selection of the winning design, and will be the only Level 1 EMD participant.

Italy, the Netherlands, and Turkey may be Level 2 participants in EMD. If they join, they will be able to exercise some degree of influence over the EMD program. Level 2 participants will have two to five representatives at the JSF Program Office.

Canada, Norway, and Denmark have been mentioned as possible Level 3 participants in EMD.

Many other countries have reportedly been invited to take part in EMD as Level 2, Level 3, or FMS participants.

As discussed in Chapter Seven, industries from virtually all the participants in the CDP are represented on the two prime contractor teams developing the demonstration vehicles.

Current Cost and Schedule

The selection of the winner of the JSF program and the beginning of EMD have been delayed about six months, to fall 2001, for a variety of reasons, including DoD's desire to examine alternative competition strategies for JSF EMD and production, and to further examine the industrial-base implications of a winner-take-all strategy. Also, there is widespread recognition that a new administration will likely want to conduct a major review of U.S. tactical fighter modernization programs before launching JSF full-scale development.

The JSF EMD is estimated to cost about $20 billion (FY94 $) and is planned to last a little over 10 years. Fourteen preproduction flight-test aircraft will be built and flown during EMD. The first aircraft to be delivered are planned to be the CTOL variant to the Air Force and the STOVL variant to the Marine Corps in 2008. The first U.S. Navy (CV) and United Kingdom (STOVL variant) aircraft are planned to be delivered in 2010.

Chapter 3

Competition During Production: Benefit or Liability?

The JSF program has been intensely competitive since 1996. The present plan calls for one of the two firms currently flight testing demonstrator aircraft, Boeing and Lockheed Martin, to be eliminated in fall 2001. The winning firm will perform all further development and manufacturing of the JSF.

This winner-take-all prospect raises the following question: Should some kind of competitive posture be retained further into the program? If so, how should it be structured? In this chapter, we examine (1) some of the special conditions that exist in defense procurement, and the consequent balance of costs and benefits created when competition is introduced; and (2) the range of possible advantages and disadvantages the government might experience from sustaining competition through production.

Special Conditions in Defense Procurement

Defense procurement operates under a set of rules and constraints that differ from the classic free-market model. Here, we briefly discuss three aspects of those differences that have a direct bearing on how competition can be introduced into defense procurement.

Not a Perfect Market

The special nature of the environment for competition in defense procurement can be illustrated in comparison with the conditions usually assumed in the economist's model of a perfectly competitive market. This comparison will help in understanding why there are inherent difficulties in introducing effective price competition into defense acquisitions. This comparison also provides a key to understanding the variety of competition-enhancing arrangements that have been developed in defense procurement. Table 3.1 lists the characteristics assumed for the perfect-market model next to the corresponding characteristics of a typical defense system acquisition in which no special arrangements have been made to introduce competition for achieving price savings or other objectives.

The conventional wisdom is that when military acquisition programs experience difficulties, expenditure is the first constraint to be relaxed, and then schedule, but that performance goals are adhered to quite rigorously. The result is frequently an increase in the unit price of the product.

The data support this description of the way capability, schedule, and price are traded off in defense acquisition (see, for example, Dews, Smith, et al., 1979). However, it is not clear that focus on this increase should be accepted as adverse criticism. The services' emphasis on high system capability is consistent with the long-established national policy that relies on capability rather than quantity for defense—a reliance that calls for the development and production of systems superior to those fielded by possible opponents. Consequently, major system acquisitions generally aim at a product that requires innovations in design and the application of advanced technologies, with all the technical uncertainty that entails.

Phases in Buyer-Contractor Relationship

Another special feature of the acquisition process for major defense systems is that it comprises several separate and distinct steps, or phases, as depicted in Figure 3.1. The buyer-contractor relationship thus involves four phases—Concept Exploration, Program Definition and Risk Reduction, Engineering and Manufacturing Development (EMD), and Production. Each phase is triggered by a milestone decision made in the Office of the Secretary of Defense (OSD).[1]

Concept Exploration. Milestone 0 approval begins the acquisition process and authorizes entry into the Concept Exploration Phase. The

[1] The most recent of the DoD 5000 series directives introduces a somewhat different phasing and nomenclature. Here, we have retained the definitions and nomenclature applied during the previous phases of the JSF.

Table 3.1
"Perfect Market" Versus Typical Market Characteristics in Major Defense System Production

"Perfect market" characteristics	Major defense system market characteristics
Many buyers and producers, none being dominant; each buyer has a choice of many producers. To a close approximation, price (a firm fixed price) is determined by the "hidden hand" of the market.	Only one buyer. Usually only one producer—the prime contractor who developed the system. Production prices (seldom truly firm fixed prices) are determined by a series of negotiations in a sole-source environment.
Product is an existing, standardized item, the same for each producer—"homogeneous"—and its characteristics are stable over time.	Product is a newly developed item, usually without close substitutes and with a design that is periodically upgraded during much of the production phase, often even afterward.
Competition focuses on price alone.	Prospective producers compete early in the development phase through "design rivalry." Buyer is concerned with product quality (especially performance), delivery schedule, and other nonprice factors. Price is not the dominant consideration in selecting the producer; quality of the product is normally given priority.
No producer has an advantage in production technology or economies of scale.	Production technology is dynamic and may differ among prime contractors and their subcontractors. Economies of scale, including cost improvement over time ("learning curve") and production-rate effects, significantly influence producer costs. A superior developer is not necessarily a more efficient producer.
The market is easy for new producers to enter.	New prime contractors seldom enter the defense sector; entrance is inhibited by the high capital investment required, the proprietary rights of others, and the administrative and contractual burdens of a highly regulated industry.
Buying the product is a simple, quickly completed, one-step transaction between the buyer and the producer, and is independent of other purchases from the same or other producers.	Acquiring a major system is a multiyear, multistep, complex process, involving scores of successive, usually interdependent, contract negotiations between buyer and producer.
The market is characterized by perfect intelligence and absence of uncertainty. Information about product price, standards of quality, number of items purchased, and delivery schedule is freely available to all concerned.	Uncertainty is a dominant, and largely unavoidable, feature. Among the market uncertainties are the • threat the system will face • most suitable system capabilities • best design approach • feasibility of development • time and other resources required to complete development and make the transition to production • deficiencies that may be revealed by operational testing.

RANDMR1362-T3.1

Milestone Decision Authority specifies the minimum set of alternatives to be examined, the lead organization, and exit criteria. In the Concept Exploration Phase, a statement of need is agreed upon. The focus of

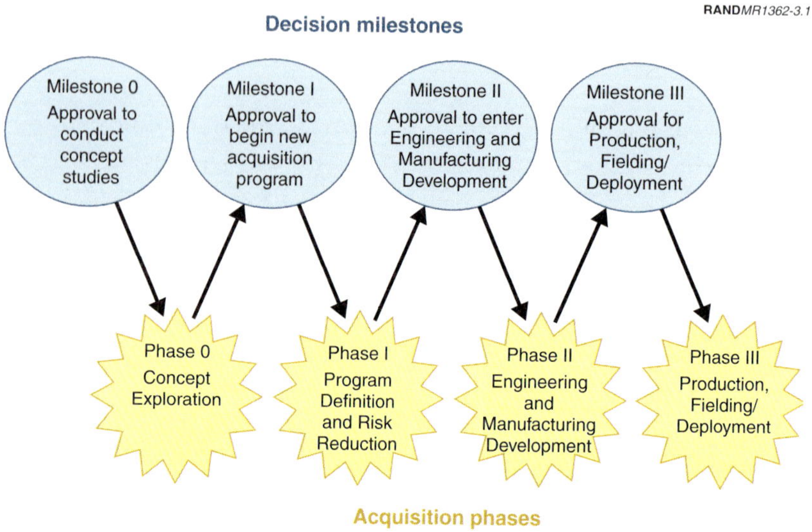

SOURCE: Joseph H. Schmall, *Introduction to Defense Acquisition Management*, Fort Belvoir, Va.: Defense Systems Management College Press, June 1996.

Figure 3.1—Acquisition Process and Phases

this phase is to define and evaluate the feasibility of alternative concepts and to provide a basis for assessing the relative merits of these concepts at the next milestone decision.

Program Definition and Risk Reduction. After a go-ahead decision has been made at Milestone I, the Program Definition and Risk Reduction Phase (Phase I) begins.[2] At this point, the acquisition strategy and concept baseline are approved. Exit criteria that must be accomplished during Phase I are established.

Phase I is characterized by measures designed to reduce the risk of incorporating new and emerging technologies. Contractors put forward designs and their feasibility is assessed; prototypes may be built and compared—"fly before buy"; preliminary estimates are made of system performance, schedule, and costs; and the trade-offs among these three variables are considered. Emphasis is on ensuring the feasibility

---

[2] Before Milestone I, some prospective prime contractors may have been informally consulted, performed special studies under contract, or submitted unsolicited proposals.

of system design and the capabilities it promises, and on reducing any inherent risks to levels deemed appropriate for moving to the next phase.

Engineering and Manufacturing Development. The next step—Engineering and Manufacturing Development—begins with a go-ahead decision at Milestone II. Proposals for EMD are requested from several contractors—almost always from the fairly small number of "primes" that participated in Phase I. Especially if prototypes have been built and tested, these proposals describe rival designs and their estimated capabilities in much more specific detail, and provide more-refined estimates of schedule and costs. In most cases, the high cost of EMD means that only a single prime is chosen in the source-selection process. The task of the EMD contractor or contractors is to bring development to the point at which transition to production can begin. EMD contracts usually call for some initial output of fully production-configured units at low production rates, which enables manufacturing and production processes to be validated. There is a heavy emphasis on testing to ensure that specifications are met and that the system is operationally effective and operationally suitable.

In EMD contractor selection, cost receives substantially more attention than it does in earlier phases. For example, although design-to-production cost goals may have been established in the request for proposals, the contractor's cost estimates for EMD and production are recognized as still being subject to much revision—typically upward. The choice of the prime contractor or contractors for EMD is normally weighted in favor of expected system capability; cost is an important but still secondary consideration.

Production. The fourth phase in this description of the buyer-contractor relationship begins with a favorable Milestone III decision to proceed to full-rate production of the system. A production contract is then negotiated, and, if there has been only a single prime contractor in the EMD phase, the negotiation is conducted in a sole-source environment. This phase often overlaps Phase II, especially when a low rate of initial production (LRIP) is a part of the program acquisition strategy. The system is produced and delivered (along with support infrastructure) to the field for operational use. To assess performance and quality, compatibility, and interoperability, Follow-on Operational Test and Evaluation may be conducted. System status is monitored to ensure that the system continues to meet the user's needs. During deployment and throughout operational support, the potential for modifications to the fielded system continues.

At the end of a system's useful life, it must be demilitarized and disposed of. During this portion of the system life cycle, the project manager must ensure the materiel requiring demilitarization is controlled. The project manager must also ensure that disposal minimizes DoD liability related to environmental, safety, security, and health issues.

Numerous Contract Negotiations

The complexity of the buyer-producer relationship is reflected in the large number of sequential contract negotiations that take place between the buyer and the prime during the course of the acquisition process. This large number of negotiations is the result of several interrelated factors (see Table 3.1), including the following:

- Very long program duration
- A widespread institutional preference for short-term, sequential decisions
- A product changing over time, with development continuing through—and usually beyond—the production phase.

For a single major acquisition, the contractual relationship between the government and the prime may continue over 20 years or more—beginning with (or even before) Phase I and ending with the last post-production upgrade or prime-contractor-handled spare-parts purchase. This decades-long buyer-contractor relationship is one reason for the numerous contract negotiations that occur.

Another reason is the widespread institutional preference on the buyer's side for short-term, sequential decisionmaking. Congress has preferred to exercise control through annual appropriations, even when multiyear contracts are approved; and most major-system contracts (or contract amendments) are negotiated for a single year's buy.

The services have compensated for uncertainties about the threat—and especially for uncertainties about future-year funding levels, allocation decisions, and the timing of new starts—by generally preferring to retain programming flexibility, avoiding long-term contractual commitments.[3]

Contract administrators and auditors prefer short commitments, which enable them to close out contract files in a few years and avoid long and complex audit trails. Compared with longer-duration contracts, short-term, quickly completed contracts may also have financial benefits for the government. For example, by shortening the contract period, the contractor receives its profits sooner and resulting taxes become payable sooner.

During EMD and the early years of production, program managers want the contractual flexibility to make desirable design trade-offs, fund major design changes, and approve at least some of the many apparently well-justified engineering change orders that are almost inevitably proposed at this time. These actions require frequent contract renegotiations or the negotiation of contract amendments. Other, often separate, contract negotiations can involve such things as long-lead-time procurements, additional test items and test support, initial and other spares, data rights, special studies, foreign assistance, new subsystems, and retrofits.

The result is that the contractual relationship between the buyer and the seller is neither the one-step transaction of the perfectly competitive market nor the four-phase transaction implied by the OSD milestone decisions.

The Benefits and Drawbacks of Competition

In a government procurement, competition can have both benefits and drawbacks, some of which are depicted in Table 3.2. The measurement of past effects of competition has produced a variety of answers, some of them conflicting. We need to carefully examine each situation, assess the likely benefits and drawbacks, and reach a judgment on the expected balance between them.

[3] However, some members of the armed services have argued for greater use of multiyear contracts as a means of "locking in" stable funding for favored acquisitions, even at the cost of decreasing the stability of other acquisitions.

Table 3.2
Primary Benefits and Drawbacks of Competition in Defense Acquisitions

Benefits	Drawbacks
Reduced prices	Additional front-end time and money needed
Staffed with best employees	Extra management complexity and effort required
Enhanced product quality through technology insertion and design refinement	Few, if any, near-term benefits
Strengthened industrial base	Outcome is uncertain

RANDMR1362-T3.2

Benefits of Competition

Most observers argue that competition produces many significant benefits (Archibald et al., 1981). Compared with a noncompetitive environment, competition results in improved product quality and lowered unit costs. They say competition forces manufacturers to quickly learn about new technologies and production techniques, fostering greater technological progress and industrial productivity. Finally, competition allows for a more equitable process under which acquisition contracts are awarded.

We do not question the value of competition as a means of inducing a firm to reduce prices. When competition or the threat of competition is perceived as real, a firm can act in a number of ways to cut costs and prices. Managers often assign their best people to a competitive program, allocate corporate capital for equipment, and fund value-engineering studies (rather than expecting the customer to fund them). A company can transfer production from an area of high labor costs, such as California or Massachusetts, to locales where labor costs are lower.

Also, management can take measures to substitute capital for labor, accelerate cost-reduction schemes, and seek out alternative vendors. A firm may be able to operate at an economical rate by producing enough parts in a few months to satisfy the contractual requirement for an entire year, then assign the workers to other tasks for the remainder of the year. In addition, a company is often able to reduce the number of engineering and manufacturing support personnel assigned to a program. Noncompetitive programs tend to be heavy in such personnel, often because the customer wants to retain the services they provide.

It is difficult to assess the effect of competition in the abstract. A contractor who needs business or is determined to increase market share acts differently than one who does not. Willis R. Greer, Jr., and Shu S. Liao (1993) discuss this "hungriness" factor; however, their capacity-utilization model uses industry-wide capacity as an input, rather than the capacity of an individual firm. Savings are contingent on speculation about what might have happened if a second contractor had not been brought into a program and assumptions about estimates of program cost without competition. If that estimate is too high, savings from competition or other causes would be easy to achieve. If it is too low, savings would be unlikely.

Improved quality assurance often is cited as a reason for second-sourcing. In some instances, the underlying reason has been a profound dissatisfaction with the initial contractor, which may be a good developer but an inefficient producer. The nature of defense procurement is such that, once a contractor is chosen to develop a major new system, the responsible military service is locked into a relationship with that contractor that could last 20 years or more. Bringing a second company into a program is an effective way to encourage greater cooperation from the initial firm.

Drawbacks of Competition

Competition also has its drawbacks and barriers. Observers note that competition requires additional time and money and entails extra management complexity and effort. Since most of the benefits of competition are long-term, not near-term, program managers have few incentives to implement competitive steps. Further, competition has uncertain and mixed results. In a risk-averse environment, this uncertainty reduces the program manager's incentive to use competition.

In discussing each drawback, we attempt to distinguish among different problems that arise during discrete phases of the acquisition cycle.

Additional Time and Money. At almost every phase in the acquisition cycle and for almost every kind of competition, adding a second competitor initially costs more than the cost of a sole source. During the Concept Exploration Phase, such funds are fairly small in absolute terms, although large relative to the overall funds available in that budget category. But competition during the Concept Exploration Phase is a well-established tradition, so funding for multiple sources is somewhat easy for a manager to obtain.

When the program moves to EMD, the magnitude of the funding required for a second, competitive source becomes large in both relative and absolute terms. Furthermore, while general statements supporting competition occur at every level in the defense establishment, this verbal support does not mean that everyone concerned with a particular program would be willing to fund competition. When the funding required to support a second, competitive source reaches the level of tens or hundreds of millions of dollars, authorization will have to come from higher up in the decisionmaking structure. Many people will have to be "sold" on the competitive action. At every level in the organization, some will be sympathetic to the request for funds, and others will see themselves as competing for the same funds. Some groups will tend to underestimate the difficulty of developing a particular system or have an interest in fielding it very quickly, and will thus resist competition during full-scale development on the grounds that it is a waste of time and money. The situation is even more complex in multiservice programs, for which each of the services must agree to put up the extra money.

When substantial amounts of money are involved, the DoD and Congress must be sold on the competition as well. When there is no great pressure for competition and when other acquisition initiatives are being emphasized, DoD and Congress can be difficult to convince. Congress tends to dislike programs with heavy front-end cost, and other, less obvious, political problems sometimes intrude (Rich, 1976). Also, funding requests are reviewed by four different congressional committees that do not automatically coordinate their decisions; so, each committee must be persuaded separately to fund a program.

Further, once funding for a competition is approved, there is no guarantee that it will be maintained. Money for competitive development programs is a prime target in a budget squeeze, and initial high-level support for competition may evaporate. Written policy supporting competition remains fairly consistent, but interest in competition changes with personnel. In the services and in the OSD, there are frequent changes in top-level personnel. When new people take over, they inevitably change priorities. The result is that it can be difficult to maintain all the funding necessary to conduct a competitive development program.

Competition can slow the program during EMD because of the time involved in testing and source selection or in qualifying a second contractor. Schedules also can lengthen because of the increased program complexity and increased bureaucratic involvement caused by competition. By lengthening schedules, competition carries the risk of raising program costs. Moreover, the risk of increased program length also is a disincentive to competition because there is usually a strong desire to deploy the system as rapidly as possible.

During the production phase, the funding required to qualify a second, competitive source appears to pose less of a problem, at least for less-complex systems or components—perhaps because, by the time the program is in production, all major conceptual issues have long since been resolved. Attention is more easily focused on the task of efficiently producing the system. Furthermore, there is some evidence of financial benefit from competitive reprocurement.

Extra Management Effort. Competition increases the workload of the Program Office. This extra work stems from two sources: planning for the competition and ensuring fairness.

If a competition is to be beneficial, extensive planning for the competitive steps is necessary. The request for proposal (RFP) must be prepared and the source-selection process designed; the Program Office must comply with certain regulations designed to ensure the fairness of the competition, which involves special security to deal with competition-sensitive material, special reports, etc.

Competition introduces the possibility of lawsuits, disputes, and charges of unfairness by contractors who lose. So, the source selection must be carried out in a way that not only chooses the best design but also raises a minimum number of questions about fairness—not an easy task. The task is complicated particularly because little information or guidance can be drawn from the experience of other programs from lessons learned reports. For the most part, program managers must plan solely on the basis of their own experience. Some program managers need no more than their past experience; others' lack of experience with the additional burden complicates planning.

Competition during production introduces still other management complications. Qualifying a second producer after production has begun can be a major effort. It is difficult and expensive to create a good technical data package (TDP) for the second contractor to use in starting production and even more difficult to persuade the first producer to pass along to a competitor the benefits of its manufacturing experience. Not all services have an in-house capability to evaluate a TDP. Without this capability, it is difficult to judge the adequacy of a TDP. Even with a good TDP, it frequently takes a major effort by the Program Office to help the second source through all its technical problems and into production.

Moreover, in developing a competitive source, the Program Office must work with both contractors on such things as quality control and configuration management. It is generally quite difficult to get two contractors to produce systems and components with interchangeable parts. If they do not do so, the Program Office faces additional problems in spare-parts procurement and logistics. Further, each added production line means an additional set of nonrecurring costs whenever there is an engineering change. Finally, if two production lines are created, the program manager must decide how hard to push each contractor in order to ensure the benefits of competition. A manager who pushes too hard runs the risk of driving one of the contractors out of the program.

As we have seen, one factor in program managers' reluctance to introduce competition is the perception that it will make management of their program more difficult and increase their workload. Since very few program managers believe that they have enough well-qualified people to cover the work of monitoring one source, they are reluctant to take on even more work or to complicate matters.

It should be noted that, under special circumstances, competition can reduce the management workload of a Program Office. Under a fixed-price development program for which the prime contractor is obligated only to a "best effort," the program manager can adopt a largely "hands-off" management style, with competition substituting for a host of conventional Program Office management controls over the contractors. This acquisition form is rarely used, because the services prefer to retain substantial control over contractor actions, even with the attendant management workload. Another possibility is for the prime contractor to act as the agent for the government in organizing competition for stipulated subsystems or components, thus relieving the Program Office of most of the burden of managing competition.

Few High-Confidence, Near-Term Benefits. Disincentives of the type described above tend to limit a manager's enthusiasm for introducing competition. The costs of competition are short-term and clear; the benefits of competition are long-term and uncertain. The incentive structure is more likely to motivate the program manager to look for strategies that return short-term benefits. Apart from exhortations in policy documents and the conventional wisdom that competition is good for everyone, few direct incentives for introducing competitive practices exist. A program manager is unlikely to be rewarded merely for introducing competition. Real cost reductions are difficult to prove and can be masked by other factors, such as inflation. Moreover, given a typical tenure of only about three years, a program manager is unlikely to be around to receive the credit for any benefits that finally accrue.

In many cases, competition is seen as impractical. There may be few or no contractors qualified to participate in a competition, and many of them may not wish to compete. Contractors often find that the uncertainties about how a competition will come out and about the criteria to be used in the source selection are sufficient to deter them from entering a competition. Qualifying a second source can be seen as impractical, because the production run is too small, the tooling for the second production line is too expensive, or the design is too complex to be transferable.

Uncertain Outcome. Retrospective studies of second-source procurement programs have not been conclusive about the effects of competition, partly because their conclusions depend heavily on the analytic methods used. Archibald and colleagues' 1981 RAND study of the Shillelagh missile, for example, showed that analysts—by using different analytic procedures—could produce vastly different estimates of the effect that second-sourcing had on procurement costs. Those estimates ranged from a cost savings of 79 percent to a cost increase of 14 percent. Some uncertainty is inevitable: If two sources are used, one cannot know the cost that would have been incurred with only a single source. That cost must be estimated and compared with the actual cost incurred through second-source procurement.

A follow-on to the 1981 report (Birkler et al., 1990) describes five methods of estimating the cost benefits of second-sourcing. For each method, the research team estimated the hypothetical single-source cost for four air-to-air missile programs (AIM-7F, -7M, -9L, -9M). For none of these programs was any of the five analytic methods unanimous in indicating that a net cost savings accrued to the government through competition. However, three of the five methods did show a net savings for two programs (AIM-7F and AIM-9L), and four methods did show a cost increase for one procurement (the AIM-7M). In any case, it is exceedingly difficult to determine a distinct cost benefit for competition. The path not taken is always an educated guess. When the results are very sensitive to the assumptions made, one must be cautious in drawing any conclusions.

Chapter 4
Analysis Procedure

The primary emphasis in this study was on ways in which competition could be retained throughout the production phase of the JSF and on whether such extended competition might prove advantageous. Our approach to this issue is outlined below, with details and results reported in Chapters Five through Ten. We also recognize that competition is not the only management option available for controlling program costs. In the second part of this chapter, we briefly discuss other options to supplement the analysis of competition.

Analysis of Competition Strategies

In this study, we examined competition in the JSF program from the perspective of a defense policymaker, identifying the information that would be required and the kinds of trade-offs that would be encountered before informed policy decisions could be made.

Our first step in designing an analysis process was to outline the different ways in which competition can be introduced into the JSF acquisition program. In this study, we identified four general categories of competition, organized as shown in Figure 4.1. The most direct form of competition—one within the JSF program itself—is to have two or more suppliers concurrently producing the system or some elements of the system. A second form of competition—one outside the JSF program—is making available other similar systems that could be substituted for the item in question and that could perform at least most of the same military missions. For example, the later variants of the F/A-18 might be considered a competitor to the Navy variant of the JSF.

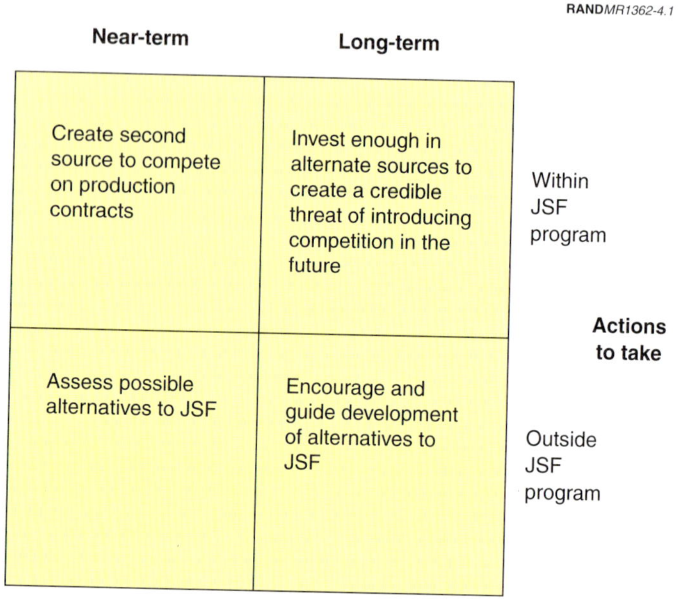

Figure 4.1—Near-Term and Long-Term Competition Options Within and Outside the JSF Program

Actions can be taken to implement these options in the near term (for example, at the start of EMD) or at a later time (for example, following LRIP or when an advanced series is introduced). In this study, we devoted most of our attention to near-term competition for production within the JSF program; we examined the other three options in lesser detail (see Chapter Nine).

Options for Near-Term Competition in Production

Introducing a second source during production of a weapon system is expected to have a variety of benefits and liabilities. The balance between benefits and liabilities can be judged by several different criteria. How the outcome of the balance is examined and evaluated differs from one criterion to another. Thus, the overall evaluation of where and whether to introduce a second production source entails different analyses and the subjective combination of the individual measures.

In this JSF study, we elected to examine the desirability of introducing near-term competition during production on the basis of two sets of criteria:

- Would competition be likely to reduce total ownership cost?
- Would competition be likely to have other benefits that cannot be easily expressed in terms of dollar cost?

Our approach to each analysis is outlined below.

Is Competition Likely to Reduce Cost of Ownership?

We first examined the issue of whether competition during EMD is likely to reduce acquisition cost. The most direct approach would be to estimate the cost for a sole-source producer, then to estimate the cost for a pair of competitive producers, and then to compare the two estimates. The first step is straightforward and draws on well-established historical data on aircraft production costs. Unfortunately, we have no comparable data or cost-estimating relationships that would enable us to estimate production costs in a competitive environment.[1]

What we do have is a limited amount of historical data showing the amount by which production cost changed when competition was introduced into an ongoing sole-source production program. To use this information, we modified a strategy suggested by Margolis, Bonesteele, and Wilson (1985):

- Estimate the expected cost of producing the JSF by a sole source, using JSF system characteristics and available cost-estimating procedures.
- Estimate the additional costs of introducing a competitor, including the front-end investment plus the inefficiencies caused by each producer's having a decreased production run.
- Determine the percentage reduction in sole-source cost that would be required to offset the additional costs of introducing competition.

[1] There have been no cases of competitive aircraft production in the United States since World War II. A few parallel production lines have been established, but their goal has been to accelerate production, not to save costs.

- Look to historical evidence on the distribution of production-cost changes when competition was introduced as a basis for estimating the likelihood of achieving competition-induced cost reductions large enough to offset the additional costs incurred by introducing a second producer.

The result is an estimate of the likelihood that the government would break even on the introduction of a competitive second source for producing the JSF. We do not estimate the "dollars saved" or the "dollars lost" as a result of introducing competition; instead, we estimate the likelihood that the overall acquisition cost to the government would be about the same for either strategy. If the likelihood is high, the government might reasonably elect to introduce competition in the expectation of achieving other potential benefits (see Chapter Seven). Likewise, a low expectation of breaking even on production cost would discourage the government from introducing competition, because the net dollar cost of production might increase enough to outweigh other possible benefits. The details of the analysis process, and the results, are presented in Chapter Five.

There is a plausible basis for expecting that, under some circumstances, a competitive production posture might lead to a higher-quality product. This higher quality would, in turn, be reflected in higher reliability and, thus, lower support cost throughout the operating life of the JSF. The available historical data do not permit a quantitative link between competitive production and product reliability. Therefore, we were unable to estimate the magnitude of expected savings in dollar costs. Instead, we assumed various levels of reliability improvement and estimated the likely consequence in reduced operations and support (O&S) costs and then determined whether such reasonably expected savings from competitive production would alter the likelihood of breaking even on introducing a competitive production source. The exact process and results are described in Chapter Six.

Might Other Benefits Be Expected from Competitive Production?

The main emphasis of this study was on estimating the likelihood of breaking even on the cost of production by introducing a competitive second source. However, we would be remiss if we ignored other possible consequences of competition. We made a cursory examination of five such consequences:

- Would competition tend to mitigate the amount of cost growth that might be expected during development and production?
- Would competition strengthen the industrial base?
- Would competition tend to reduce overall risk levels in the program?
- Would competition enhance the rate of introducing technical innovations and the overall quality level of the product?
- Would competition affect the kind and scope of international participation?

We included the results in our overall evaluation of competition during production. The first consideration lent itself to a quantitative examination. We pursued the others qualitatively, and the results of these secondary examinations are presented in Chapter Seven.

Tools Other Than Competition to Control JSF Costs

The focus of this study is on the effects of competition as a means of controlling costs in the JSF program. Other tools are also available to government program managers and other acquisition officials to help control costs, even in a winner-take-all development and production scenario. We summarize those tools here to supplement the analysis of competition discussed throughout this report.

Institute a Vigorous Cost-Performance Trade-Off Program Among All Stakeholders

Key to controlling costs on the JSF is to have cost and performance data available to all stakeholders so they have a clear understanding of where marginal improvements or decreases in performance begin to drive nonlinear increases or decreases in cost. This kind of information is the basis for the DoD policy on Cost As an Independent Variable (CAIV). With better understanding of the costs and performance, the importance and influence of the requirements can be better understood. Those requirements that make little difference in operational performance but that inordinately drive costs can be modified so that only well-analyzed requirements are met by JSF performance. Clearly, cost analysis efforts will have to focus on these areas to ensure that program performance, costs, and risk margins are well understood.

Provide Award Fees as Incentives to Improve Producibility and Reliability During EMD and Production

Award fees are a powerful incentive to ensure that contractors are especially focused on areas that are important to DoD and that drive costs. However, such fees are not always easy to design. While production contracts are normally firm fixed-price instruments, the basis of the negotiated prices is usually the costs incurred to produce previous lots of aircraft, adjusted for cost improvement—learning curves—with a profit percentage added. Contractors may actually lose revenues through producibility initiatives that lower their costs and, therefore, their profits. In addition, costs to produce an aircraft are often locked in early in aircraft design. Producibility improvements may be more difficult to introduce once a design is finalized. Therefore, it is important for DoD to foster innovation during EMD and production by awarding fees that are tied to well-demonstrated initiatives that will reduce JSF production costs. This will require some agreement between DoD and the contractors on a production baseline cost, so that initiatives can be shown to reduce costs from that baseline.

Improvements in reliability are even more difficult to demonstrate, particularly since the effects (and the cost savings) are well downstream of design and production activities. Unless the contractors have some specific incentive to design reliability in and demonstrate these improvements through reliability-growth curves during EMD, they have little financial incentive to save operations and support costs after equipment is introduced into the operational forces. Many suggest that, if the contractor is assured that it will receive the contract for follow-on support, there is enough incentive to try to aggressively improve reliability during design. With no contracting relationship and a period from design to operational support of 10 years or more, this incentive seems tenuous at best. However, if improvements in reliability can be designed and demonstrated through well-analyzed and tested reliability-improvement projections during EMD, contractors can be motivated more successfully with immediate financial rewards than with assurances of getting operational support business.

Require an Active Production Price Commitment Curve Following a Major EMD Milestone

Many targets, objectives, and promises concerning costs are formulated during aircraft development, and many of these tend to either increase inexplicably over time or fall victim to the cost-based negotiations addressed above. Establishing an active, mutually agreed-upon price commitment curve following a key milestone during EMD (for example, following Critical Design Review) would appear to benefit both DoD and the contractor.

The benefit to DoD would be a known price, which the services could use to program and budget for aircraft procurement year to year (or during the multiyear period). A commitment to price would motivate the contractor to focus on producibility continuously but would also allow it to keep the savings from these cost-reduction initiatives on a dollar-for-dollar basis, since the contractor would sell aircraft for the agreed-upon price, regardless of its cost to produce.

Clearly, over time, excess profits could become a concern, so there would have to be periodic renegotiations on the basis of the actual costs, which could be done for each multiyear contract, normally every five years. Thus, instead of losing profit on each dollar saved in production costs, the contractor would achieve a dollar of profit for each dollar saved. A clear designation of the price of the JSF—with the contractor prepared to stand behind its price, much as the commercial aircraft producers must do—has definite advantages for convincing Congress and the American people that the JSF is not going to become another program plagued by cost overruns. Clearly, requirements must be controlled in order to avoid "opening Pandora's Box" and price renegotiations with each JSF lot.

Emphasize Warranties as Part of EMD and Production Phases

Like the commitment to production prices, the frequent promise of system reliability is often not achieved after aircraft are fielded. As discussed above, award fees can motivate contractors to design in reliability during EMD, and analytically derived reliability-improvement curves can monitor reliability during testing. But design reliability can be fully proven only after a system is fielded. Unfortunately, the contractual arrangements under which DoD must procure its aircraft mean that award fees for achieved reliability can be paid only over the life of the contract, which ends shortly after aircraft delivery. So success in achieving reliability after an aircraft is designed and produced does not result in financial reward to the contractor.

A warranty may be the answer. A warranty of reliability for any number of years can be part of the production contract. This mechanism has been successfully implemented on missiles (which spend most of their operational life in storage). However, unlike the commercial aircraft market, in which warranties are commonplace, DoD has not had an aircraft (particularly a fighter) covered by a warranty. One of the arguments against the use of warranties on combat aircraft is the fact that the actual operational usage is more stressing than the designed usage, so the contractors would object to this sort of "blank check" situation. With the incorporation of the Prognostic Health Monitoring (PHM) system planned for the JSF, this situation can be alleviated. Aircraft usage above the designed limits could be analyzed and warranties either adjusted or voided. Health monitoring on commercial engines is a prime example of how the commercial engine manufacturers can analyze an engine's usage over its life and warranties can be adjudicated with airlines. Such monitoring could be applied to the JSF.

Chapter 5
Analyzing Acquisition Costs

In the analysis below, we consider introducing a second contractor early in the JSF program. Our focus is on what makes sense at the beginning of EMD, which should occur approximately a year after this study. For this study, we have cost data regarding the baseline winner-take-all program from both of the JSF contractors plus the JSF Program Office. The alternative is to introduce competition, for which there is no information.

Analyzing the effect of competition on acquisition costs is always complicated by the absence of data regarding the "path not taken" or "the path for which there is no data." The baseline is no competition, and the alternative is introducing a second contractor at some point during the acquisition process. The second contractor will produce some number of units during the completion of the procurement program. Our approach employs a modification of a break-even analysis technique that was developed several years ago specifically to handle "the path for which there is no data." The next section describes this approach. The section following it presents analytic results for several possible scenarios for implementing competition in the JSF program. The final section presents the findings of several studies regarding production cost savings, or losses, on past competitive programs and discusses how these results can be used to assess the likelihood that the JSF may be able to achieve savings under competition.

Break-even Analysis

There is no reliable, quantitative method for estimating the magnitude of cost savings that may occur when a second source is introduced in a production program. To make such an estimate requires knowledge of, or assumptions regarding, the behavior of both the prime contractor and the second source under competition. What are their total business bases? What are their attitudes toward risk? Are they willing and able to reduce engineering or indirect staff to reduce costs? These and many other questions would need to be answered.

We can, however, estimate the additional costs of implementing a second contractor and then deduce the recurring production savings needed to offset the additional cost. The implementation of this approach is a modification to the break-even analysis suggested by OSD/PA&E in 1985 (Margolis et al., 1985). *Breakeven* is the condition in which the cost to the government using two contractors is equal to the cost of using only the prime contractor. It is expressed by the relationship

$$TC_{ss}(Q,R) + INV_{ss}(R) = TC_1(q_1,r_1) + TC_2(q_2,r_2) + INV_c(r_1,r_2) \quad (5.1)$$

where

$TC_{ss}(Q,R)$ = recurring cost of the single-source contractor to produce quantity Q at peak rate R

$INV_{ss}(R)$ = nonrecurring cost to bring the single-source contractor to peak rate R

$TC_1(q_1,r_1)$ = recurring cost for (competitive) contractor 1 to produce quantity q_1 at peak rate r_1

$TC_2(q_2,r_2)$ = recurring cost for (competitive) contractor 2 to produce quantity q_2 at peak rate r_2

$INV_c(r_1,r_2)$ = nonrecurring cost required to bring both contractors to peak production rates of r_1 and r_2.

This definition of the break-even model is valid for the special case where the entire production quantity is either supplied by one source or split between two competing sources, with no production occurring before the quantities examined in the analysis. Therefore,

$$Q = q_1 + q_2$$

We can rewrite Equation (5.1) as

$$TC_1(q_1,r_1) + TC_2(q_2,r_2) + INV_c(r_1,r_2) - TC_{ss}(Q,R) - INV_{ss}(R) = 0 \quad (5.2)$$

If the two competing contractors behave as the prime would as a sole source, then Equation (5.2) is expected to be greater than zero. The quantity INV_c will be greater than INV_{ss} because additional costs are incurred to establish the second contractor: The second source must have sufficient tooling, it must be qualified to produce the end item, it has its own set of overheads, and so on. The magnitude of INV_c will vary as a function of the competition design. If the competition is designed to allow for either contractor to win 100 percent of an annual buy quantity, the INV_c would be approximately double the INV_{SS}. If the two contractors follow the same production-cost improvement curve[1] as the sole source does, the production costs for the two competing contractors will be greater than the total cost for the single contractor as a sole source, because of "loss of learning." The only way that Equation (5.2) can equal zero is if the competing contractors' behaviors change to reduce the sum of the first two terms below what would be obtained if they followed the sole source's production-cost improvement curve.

For the present study, we requested estimates from both prime contractors, as well as from the JSF Program Office, regarding the two sole-source teams. These data provide the basis for determining TC_{SS} and INV_{SS}, and thus the cost of establishing the dual-source arrangement using the nonrecurring cost data for the sole source. We then calculated the ratio of the left side of Equation (5.2) to the total production cost for the sole source to produce the entire quantity—which equals the percentage by which the net cost of the two sources exceeds the sole-source production cost when the two sources do not change their production-cost behaviors relative to that of the sole source. Expressed another way, the required cost reduction (*RCR*) is the percentage decrease relative to the sole source's production cost that must

[1] According to Harold Asher (1956): "The theory of the progress curve in its most popular form states that as the total quantity of units produced doubles, the cost per unit declines by some constant percentage." The cost per unit can be interpreted as either the cumulative average cost or the cost of a specific unit. In the present study, we use the specific-unit interpretation. Hence, for an improvement curve with a 90-percent slope, the cost of the second unit is 90 percent of the cost of the first unit. The cost of the 200th unit is 90 percent of the cost of the 100th unit. And so on.

be achieved for the two competitors to offset the loss of learning and the additional start-up costs of the second firm:

$$RCR = \frac{TC_1(q_1, r_1) + TC_2(q_2, r_2) - TC_{SS}(Q, R) + INV_C(r_1, r_2) - INV_{SS}(R)}{TC_{SS}(Q, R)} \quad (5.3)$$

For simplicity, we refer to this expression as *RCR* in the remainder of this report. Equation (5.3) yields the *net savings* required to cover both the loss of learning and the investment costs of establishing the dual production sources. If the two *INV* terms are deleted from the expression, the result is the *gross savings*, or the recurring cost savings.

Required Cost Reduction for the JSF Program

Competition in the production phase of a weapon system can be structured in several ways. We first outline the specific cases we examined and the related assumptions we made, then we calculate *RCR* values for each case.

Cases Examined

Two broad second-source strategies are outlined in Table 5.1. We assume that a basic design has been established by the prime contractor, that a series of vendors has been established for the major subsystems (landing gear, ejection seat, gun, etc.) and for the mission system components, and that the prime has set up a factory for final assembly and check-out (FACO). Each of these activities can also be performed by a second source. In principle, that second source can produce the various components and functions in either of two ways: build the system element using the exact design created by the prime contractor or the prime's vendor (Build-to-Print, or BTP) or design and manufacture its own variant of the system element so that it can be directly integrated into the overall system design (Form-Fit-Function, FFF).

Which of the different second-source strategies is more or less appropriate and widely applied varies with the system element, as indicated by the symbols in the table. For major structure assemblies, it is most common for the second source to build to print; it is generally impractical to have different structure designs intermixed in a vehicle.

Table 5.1
Options for Second-Source Participation During Production

System element	Method employed by second source	
	Build-to-Print	Form-Fit-Function
Airframe structure	+	–
Major subsystems	+	+
Mission system components	–	+
Final assembly and check-out	?	?

RANDMR1362-T5.1

NOTE: + = more appropriate strategy
– = less appropriate strategy
? = could be similar or different.

However, for many subsystems and for mission system components, it is sometimes practical and desirable for the second source to create its own design in such a way that it can be inserted into the basic vehicle and function just as does the original (prime's vendor) design. Finally, if the second source also sets up a production line for final system assembly and check-out, the tooling and testing methods might be similar or different from that of the prime, depending on characteristics of the system.

It is important to distinguish between BTP and FFF, because the cost of establishing a second source is significantly different under the two strategies. The additional design, development, and test efforts required in the FFF option usually incur greater start-up cost. Sometimes those additional costs are justified because the effort introduces lower fabrication costs or achieves improved performance over that of the original design.

Data provided by Boeing, Lockheed Martin, and the JSF Program Office, combined with data available within RAND, permitted us to analyze each of the elements shown in Table 5.2 for both a Build-to-Print option and a Form-Fit-Function option.

We also analyzed the complete mission system suite and the complete airframe (including subsystems). The JSF engine is already under competition and is not addressed in this study.

Table 5.2
Airframe and Mission System Components in RAND Study

Airframe structure	Airframe subsystems	Mission system components
Forward fuselage	Landing gear	Radar
Center fuselage	Ejection seat	Communications/navigation/ identification
Aft fuselage		Electronic warfare
Empennage		Distributed infrared aperture system
Control surfaces and edges		Targeting forward-looking infrared
Wing		Integrated core processor
		Controls and displays
		Vehicle management system
		Stores management system
Final assembly and check-out		

RANDMR1362-T5.2

Analysis Rules and Assumptions

For Build-to-Print, we assumed that only one design is produced by both contractors. For the Form-Fit-Function calculations, we assumed that each contractor could provide its own design, which could be integrated into the total weapon system and satisfy all weapon system requirements.[2] For all analyses, we assumed that the quantity Q is divided evenly between the two competing contractors ($q_1 = q_2 = Q/2$),[3] that the competitive contractors have the facilities to produce at a peak rate that is two-thirds of the sole-source peak rate ($r_1 = r_2 = \frac{2}{3}R$), and that competition begins with the first LRIP units produced following EMD.[4] For all analyses, costs are in constant FY94 dollars (JSF program base year).

The baseline sole-source costs were developed from the JSF Program Office's assessment of the contractors' costs to produce 3,002 aircraft. The JSF Program Office provided estimates for functional cost elements, labor, and material. RAND researchers adjusted these estimates to create a set of composite aircraft costs.

---

2 While this is not a realistic assumption for many of the cases examined (see Table 5.1), it provides an extreme boundary for the costs of establishing the second source.

3 A 50:50 split was selected because it results in the greatest loss of learning.

4 The recurring-cost calculations do not account for production-rate effects. The nonrecurring costs for the two competing contractors are determined from a peak rate of two-thirds of the sole-source rate.

Assumptions for aircraft and mission-system-equipment quantities produced during EMD are shown in Table 5.3. The "Equivlent number built during EMD" column does not necessarily indicate the number of complete systems produced; the number is used to determine the cost-improvement effects achieved during production of items and components during EMD. The right-hand column shows the slopes of the cost-improvement curves used in the break-even calculations.

For all airframe sections and variants, we assumed cost-improvement curve slopes to be 79 percent for manufactured items and 92 percent for purchased equipment and materials. Slopes for all mission systems were developed by RAND researchers, with the ranges indicating uncertainty in the estimates. All slopes were assumed to be constant throughout the production runs (3,002 aircraft for the sole-source case and 1,501 each for the dual-source case). For FFF cases, the second source was assumed to produce the same number of EMD units as the prime. Both start production at the same improvement-curve position. For BTP cases, some learning is assumed to transfer from the

Table 5.3
EMD Equivalent Quantities and Cost-Improvement Curve Slopes Used in the Analysis

System	Item	Equivalent number built during EMD	RAND cost-improvement curve slope (percent)
Airframe structure	Conventional takeoff and landing airframe	6	79
	Carrier variant airframe	5	79
	Short takeoff/vertical landing airframe	5	79
Mission system components	Radar	11	88–93
	Communications/navigation/identification	11	89–93
	Electronic warfare	11	89–93
	Distributed infrared aperture system	11	85–91
	Targeting forward-looking infrared	11	85–91
	Integrated core processor	21	90–95
	Controls and displays	21	85–93
	Vehicle management system	21	88–93
	Stores management system	16	89–93

RAND*MR1362-T5.3*

prime, plus the second source builds two qualification units. Both competitors are assumed to start production from the same point on the cost-improvement curve for each component or section.

For FFF cases, we assumed that the second source's EMD costs equal the prime's EMD costs.

For BTP cases, we made a more detailed estimate for second-source costs:

- For mission system components, the second-source nonrecurring hardware development cost was 50 percent of the prime's cost and software development cost was 20 percent of the prime's software development cost.
- For airframe sections, the second-source EMD costs were estimated as shown in Table 5.4.

We assumed the prime would build 12 aircraft in EMD (the quantities shown in Table 5.3 do not represent complete aircraft).

Rate tooling costs were determined by allocating the JSF Program Office's estimate for total rate tooling cost ($1,500M) between airframe ($1,200M) and mission systems ($300M). This allocation is based on F/A-18E/F actuals. We further allocated these values to sub-elements, using contractor data.

JSF Cost Savings Needed

The *RCR* values for the airframe sections for BTP and FFF are shown in Table 5.5. The comparable results for the mission system components are shown in Table 5.6. The ranges in Table 5.6 reflect the improvement-curve ranges shown in Table 5.3.

To facilitate comparison with the historical experience discussed in the following section, the values in these figures are based on undiscounted costs.

Past Experience with Introducing Competition in Production

To evaluate the likelihood of achieving the required cost reductions, we turned to historical experience. Several studies of competition in procurement have been conducted over the past 30 years, with the most re-

Table 5.4
Estimating Assumptions for Second-Source Airframe EMD Costs Under BTP Scenario

Cost element	Factor (percent of prime's costs)	Rationale
Nonrecurring engineering	25	Covers translation of production methods, process, etc., plus representation on appropriate Integrated Product Teams.
Nonrecurring tooling and tooling quality control	34	Facilitates both contractors' producing at 2/3 of the sole-source peak annual rate.
Subcontract	10	Covers design translation for vendors, plus representation on appropriate IPTs.
Nonrecurring purchased equipment	19	Equates to first 2 of 12 units (2/12), assuming 92% slope.
System test	26	Equates to first 2 of 12 units, assuming 79% slope.
Ground test	5	Assumes second source participates in IPTs, but its 2 test articles do not require static and fatigue testing
Mock-ups	0	Assumes digital, 3D database
Flight test	17	Assumes 2 aircraft built for flight test and OT&E.
Operational test and evaluation	17	Assumes 2 aircraft built for flight test and OT&E.
Survivability test	0	Assumes no live-fire tests needed for BTP items.
Systems engineering/ program management	25	Assumes these costs are directly proportional to nonrecurring engineering.
Support and training	34	Facilitates both contractors' producing at 2/3 of the sole-source peak annual rate.

RAND*MR1362-T5.4*

cent completed in the early 1990s. Those historical studies cover a wide variety of weapon systems, subsystems, and components. In all of those cases, the program started with a sole-source producer, and a competitive second source was introduced later in the production run.

We reviewed reports and papers describing several dozen of those earlier studies, some of which presented data drawn from still other studies. The four-month schedule for the JSF study did not allow us time to collect and review all the original source documents and redo the cost savings analyses on a consistent basis; instead, we were limited to using the available studies and analyzing the results they presented.

Table 5.5
Airframe Component Break-even Estimates

Airframe element	BTP (percent)	FFF (percent)
Forward fuselage	31	51
Center fuselage	30	46
Aft fuselage	28	39
Empennage	30	49
Wing	32	54
Edges	27	33
Landing gear	30	45
Ejection seat	31	51
Final assembly and check-out	26	27
Complete airframe	30	46

RAND*MR1362-T5.5*

Table 5.6
Mission System Component Break-even Estimates

Mission system component	BTP (percent)	FFF (percent)
Radar	14–20	23–29
Communications/navigation/ identification system	15–20	26–31
Electronic warfare system	12–16	18–22
Integrated core processor	16–21	34–40
Distributed infrared aperture system	14–22	23–30
Electro-optical targeting system	19–27	34–41
Controls and displays	19–23	34–38
Stores management system	15–20	30–35
Vehicle management system	14–18	28–32
Complete mission system	14–20	25–31

RAND*MR1362-T5.6*

We found that a variety of analytical methods was used in the studies we reviewed; in many cases, the results from one study are not consistent with those from another study. Some compared the first competitive buy to the last noncompetitive buy. Some compared the competitive buys to the total of all buys. Some compared the competitive buy to a projection of the noncompetitive experience for the same quantity. Some savings were calculated only for the completed buys; other savings were projected to the completion of the program (as envisioned at the time of the study). Some used cumulative average cost-improvement curves, and others used unit cost-improvement curves. A couple of the studies made adjustments for economies of scale (capacity utilization, production rate). A few used discounting, but the rates were not always specified. Some studies calculated gross savings, and others included nonrecurring costs for establishing the second source, but not always the *same* set of nonrecurring costs. The documentation was not always clear regarding these points.

We also found important variations in the programmatic backgrounds of the competitive buys examined in the studies. Competition was implemented in several programs because the customer was not satisfied with the original contractor's performance. In some programs, cost was believed to be too high, but there were also instances of poor product quality or reliability. Also, the programs differed in the timing of the start of competition. Numbers are not available for all programs, but for 20 competitive split-buy programs, the quantity produced by the prime contractor prior to competition ranged from a low of 4 percent of the total to a high of 87 percent of the total.

Of the many sources we reviewed, we were able to obtain data from six that appear to be methodologically consistent: Daly et al., 1979; Drinnon and Hiller, 1979; Beltramo, 1983; Kratz et al., 1984; Flynn and Herrin, 1989; Birkler et al., 1990. The savings are based on actual costs or projections to the end of the program (not first competitive lot compared with the last noncompetitive lot). The data are all undiscounted and are based on gross savings (i.e., accounting only for changes in recurring costs, with no accounting for investment needed to create the competitive production line).

To support our analysis of the JSF program, we treated the electronics systems and equipment separately: They are most similar to the mission system equipment, and they typically have much shallower cost-improvement slopes than do major hardware components. Results are shown in Table 5.7. We grouped all other historical data together for use when examining the airframe elements, with results shown in Table 5.8. Most of the non-electronic items are ships and missiles; there

Table 5.7
Estimated Cost Savings (percent) from Competition in Electronics Systems and Equipment Programs

Equipment	Low	High	Count
AN/ARC-131 Radio	–16.1	–16.1	1
SPA-66 Radar indicator	–3.4	–3.4	1
PP-4763/GRC Power supply	0.5	0.5	1
Aerno 60-6042 Elec. cont. amp.	8.5	8.5	1
AN/ASN-43	10.7	10.7	1
UPM-98 Test set	11.5	11.5	1
FAAR Radar	16.6	16.6	1
FAAR TADDS	18.2	18.2	1
AN/SGS 23 208A Transducer	32.3	32.3	1
PRT-4	42.3	42.3	1
AN/FYC 8X	43.2	43.2	1
AN/ARA-63 Radio receiver	57.9	57.9	1
FGC-20 Teletype	4.0	39.9	3
Aerno 42-2028 Generator	7.3	19.9	2
APX-72 Airborne transponder	9.4	27.1	3
SPA-25 Radar indicator	10.7	48.8	3
AN/GRC-103	11.9	60.1	3
TD-660 Multiplexer	14.2	38.3	3
AN/PRC-77 Manpack radio	20.5	41.9	3
MD-522 Modulator	25.9	58.6	3
Aerno 42-0750 Voltage regulator	29.2	54.8	2
TD-204 Cable combiner	35.5	62.1	3
TD-352 Multiplexer	36.0	58.0	3
U.S. M-181 Telephone test set	36.3	56.0	2
TD-202 Radio combiner	36.5	46.8	3
CV-1548 Signal converter	40.2	64.0	3
AN/GRC-106	41.8	43.3	2
60-6402 Electric control	49.4	52.7	2
AN/ARC-54	55.0	63.1	2
MK-980/PPS-5	56.0	66.5	2
AN/APM-123	61.2	67.7	2

RAND*MR1362-T5.7*

Table 5.8
Estimated Cost Savings (percent) from Competition in Missiles, Ships, and Related Programs

Missiles, ships, etc.	Low	High	Count
Dragon—round	2.8	2.8	1
F404	5.1	5.1	1
TAO 187	5.1	5.1	1
LCAC	8.2	8.2	1
VLS canisters	9.4	9.4	1
AIM-54C—G&C	11.0	11.0	1
Dragon—tracker	12.3	12.3	1
VLS launcher	16.1	16.1	1
Mk 48 torpedo	16.3	16.3	1
CG 47	19.6	19.6	1
Std Missile 2—G&C	20.6	20.6	1
Tomahawk	20.7	20.7	1
Std Missile 2—motor	23.9	23.9	1
LSD 41	28.3	28.3	1
Bullpup—Martin	31.7	31.7	1
Std Missile 2 (RIM-67A)	34.0	34.0	1
AIM-9D/G	–71.3	0.7	4
Mk 46 torpedo	–36.4	–30.9	2
AIM-9M	–35.4	12.7	5
AIM-7M	–28.6	5.3	5
AIM-7F	–25.0	9.0	7
Rockeye	–23.0	25.5	4
Shillelagh	–8.0	9.4	4
Std Missile 2 (RIM-66A)	–4.2	59.2	3
AIM-9L	–3.8	24.0	5
TOW missile	8.9	26.0	5
Mk 48 torpedo—electronic assembly	11.6	47.0	2
Bullpup—Combined	18.7	26.5	4
Hawk—motor parts	19.9	49.9	3
Mk 48 torpedo—warhead	23.7	48.6	2
TOW launcher	30.2	44.2	2

RAND*MR1362-T5.8*

have been no instances since World War II in which aircraft were produced by competitive sources.

There is some overlap among these studies. Tables 5.7 and 5.8 present the highest and lowest values of percent cost savings reported in the historical documents. The "Count" column shows the number of studies that reported estimates on a particular system. If a weapon system was included in only one study, the high and low values are the same.

Scatter diagrams of the high and low savings estimates from Tables 5.7 and 5.8 are shown in Figures 5.1 and 5.2, respectively.

To assess the likelihood of obtaining different levels of savings, we counted the number of points for which the minimum savings estimates were greater than 40 percent, 30 percent, 20 percent, 10 percent, and 0 percent. The results are summarized in Table 5.9.

Relating these data to the JSF *RCR* values presented in Tables 5.5 and 5.6 required some additional insight and interpretation. Relative to

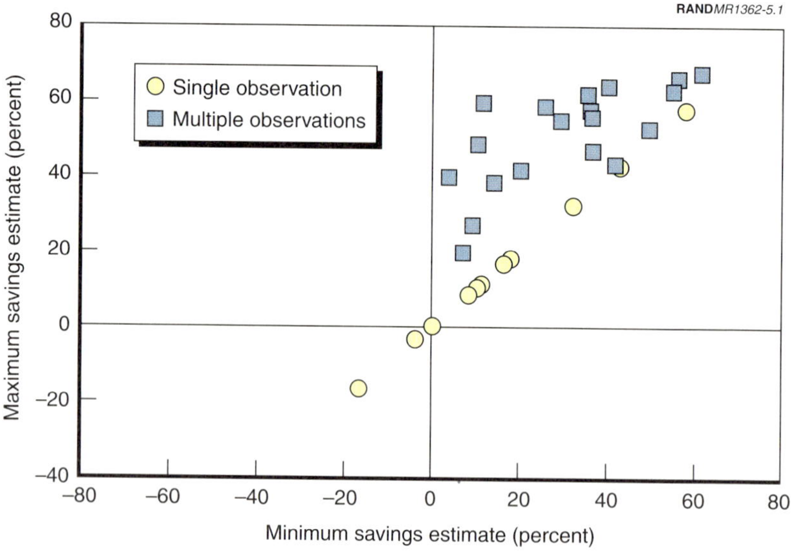

Figure 5.1—Maximum and Minimum Estimates of Savings from Competition in Electronics Programs

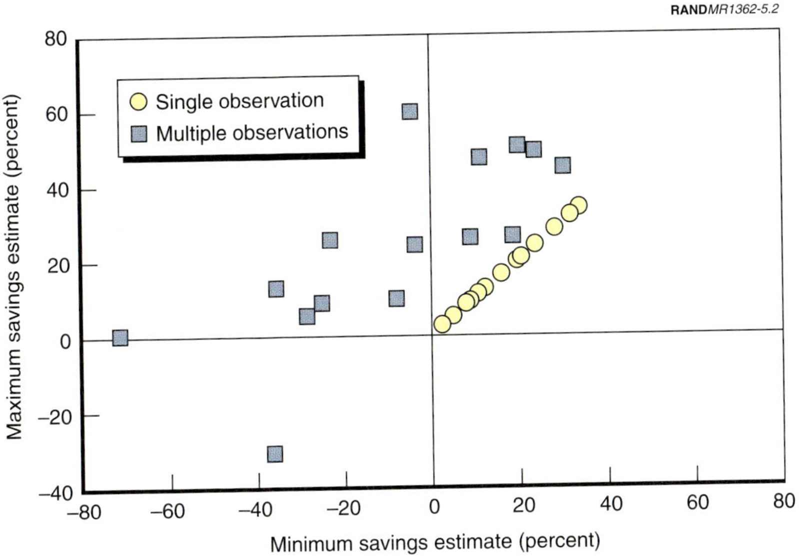

Figure 5.2—Maximum and Minimum Estimates of Savings from Competition in Missiles, Ships, and Related Programs

Table 5.9
Fraction of Programs Examined That Achieved Savings

Savings achieved (percent)	Missiles and ships	Electronics
>0	7/10	9/10
>10	5/10	8/10
>20	3/10	6/10
>30	1/10	5/10
>40	Nil	3/10

RANDMR1362-T5.9

Equation (5.3), the historical production-cost savings (*HPCS*) is represented by Equation (5.4):

$$HPCS = \frac{TC_{1A}(q_1, r_1) + TC_{2A}(q_2, r_2) - TC_{ss}(Q, R)}{TC_{ss}(Q, R)} \qquad (5.4)$$

where the subscript A indicates the values are either "actuals" or are projected from some actual data.

The differences between Equations (5.3) and (5.4) are critical to being able to use the historical data to judge the likelihood of achieving the required savings indicated by Equation (5.4). Most obviously, *Equation (5.4) does not contain any terms relating to investment costs.* As noted above, the historical studies varied in the fidelity with which they incorporated such costs. Furthermore, all the historical programs are Build-to-Print and have some amount of sole-source production prior to competition. In RAND's analyses, the second contractor participates in the program from the start of EMD, in both BTP and FFF scenarios, and begins production at the same time as the original contractor. Such participation is expected to cost more than setting up a second source after prior production by the original contractor. We disregarded investment costs in the historical data because they are not reported uniformly and they are not consistent with the basis for the required investment costs in our JSF scenario. Instead, we looked for savings, as indicated by Equation (5.4), to help offset the required JSF investment cost.

The difference between $(TC_{1A} + TC_{2A} - TC_{ss})$ in Equation (5.4) and $(TC_1 + TC_2 - TC_{ss})$ in Equation (5.3) is critical—and subtle. A zero value for *HPCS* indicates that *all recurring cost consequences* of switching to two sources have been exactly offset through actions taken by the two competing contractors.[5] These consequences include loss of learning, as well as production-rate or business-base effects on both direct and indirect costs. As used in this study, $(TC_1 + TC_2 - TC_{ss})$ in Equation (5.3) represents only the loss of learning resulting from competition. It does not include any of the other recurring-cost consequences. Thus, these terms have inconsistent definitions. Furthermore, we do not know the magnitude of the recurring-cost consequences for any of the historical programs. We have only the

[5] A negative value for *HPCS* indicates that savings beyond those required to offset recurring-cost consequences have been generated. These savings are available to offset the additional investment costs of establishing competition. A positive value for *HPCS* indicates that the contractors did not offset the recurring-cost consequences.

values for *HPCS*, not for any of the terms on the right-hand side of Equation (5.4). From the sources available to support this study, we do not even know the quantity split between the leader and the follower.

For the JSF analyses, we made no estimates for recurring-cost consequences other than loss of learning, and we calculated that, based on a 50:50 quantity split—arguably the worst case. To assess the likelihood of achieving our calculated *RCR*s, we needed to address the difference between what is inherent in the *HPCS*s and how we calculated the JSF *RCR*s. First, we assumed that the production-rate effect is the same (in percentage terms) in the historical programs and in the JSF cases and can be ignored. Then, we estimated the magnitude of the loss of learning inherent in the historical data and used the results to adjust the values in Table 5.9.

Two of the historical studies—Lovett and Norton, 1978, and Beltramo, 1983—present quantity and improvement-curve slopes that permitted calculation of loss of learning to be estimated for 20 programs. These documents provide the quantity produced before competition and the quantity produced competitively. They do not indicate the split between the contractors during competition. To estimate the loss of learning recovered by these historical programs, we assumed the split to be 50:50. We also assumed that both contractors continue on the original contractor's improvement curve. The maximum loss of learning is 15 percent, and the minimum is 0.1 percent. Incorporating the average, 6.6 percent, as a shift to the historical gross savings realized results in a slight improvement over Table 5.9 in the fractions of programs that achieved savings and covered the nominal loss of learning, as seen in Table 5.10.

Likelihood of Achieving Savings Needed to Break Even

The JSF airframe *RCR* values for the BTP option, as shown in Table 5.5, are displayed graphically in Figure 5.3. The required cost savings values cluster around 30 percent. For the airframe FFF option *RCR*s shown in Figure 5.4, only FACO is near 30 percent. The other components range from about 35 percent to about 55 percent, averaging

Table 5.10
Fraction of Programs Examined That Achieved Savings and Covered Nominal Loss of Learning

Savings achieved (percent)	Missiles and ships	Electronics
>0	8/10	10/10
>10	7/10	9/10
>20	4/10	7/10
>30	2/10	5/10
>40	Nil	4/10

RANDMR1362-T5.10

over 45 percent. The historical record indicates that roughly half of the non-electronic systems achieved slightly less than a 20-percent savings from competition. This is indicated by the blue bar in the figures. Thus, the likelihood of achieving the necessary savings for the airframe cases does not appear good.

The JSF mission system *RCR*s for the BTP option, as shown in Table 5.6, are plotted in Figure 5.5. The values range from 12 to 27 percent. For the FFF option (see Figure 5.6), they range from a low of 18 percent to a high of 41 percent. These values are lower than the airframe values primarily because the cost-improvement slopes for electronics equipment are shallower than for airframe elements. Table 5.10 indicates that approximately half of the electronics equipment systems achieved a 30-percent savings.

The mission system cases appear to be much more favorable for competition. However, the analyses presented in this study are all based on the projection of a production run of 3,002 identical aircraft (excepting the three service variants). Hence, each of the competing contractors has 1,501 aircraft over which to achieve sufficient savings to recover the extra investment costs plus the loss of learning.

But the pace of evolution of avionics technology is much greater than that for airframe technology. Although there will be inevitable changes to the airframe during the total JSF production run, the magnitude of change will be modest compared with the entire airframe pro-

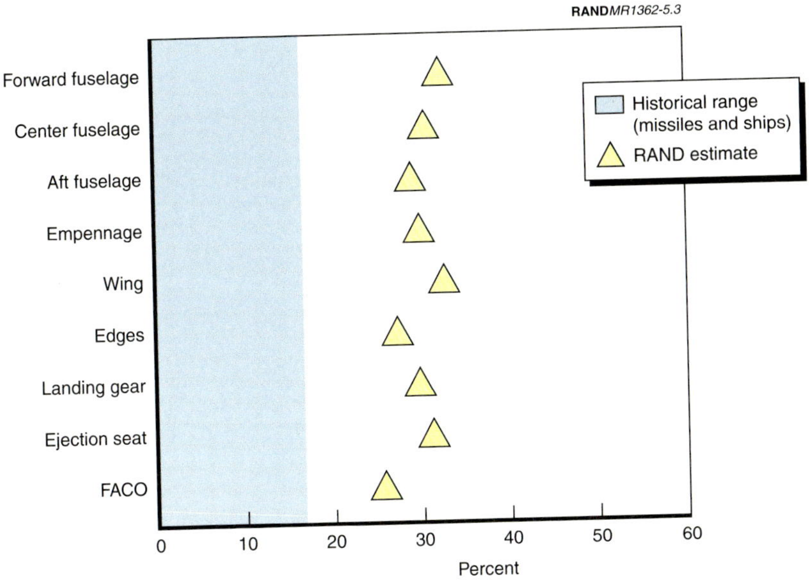

Figure 5.3—Airframe Component Break-even Estimates (Build-to-Print) from Table 5.5

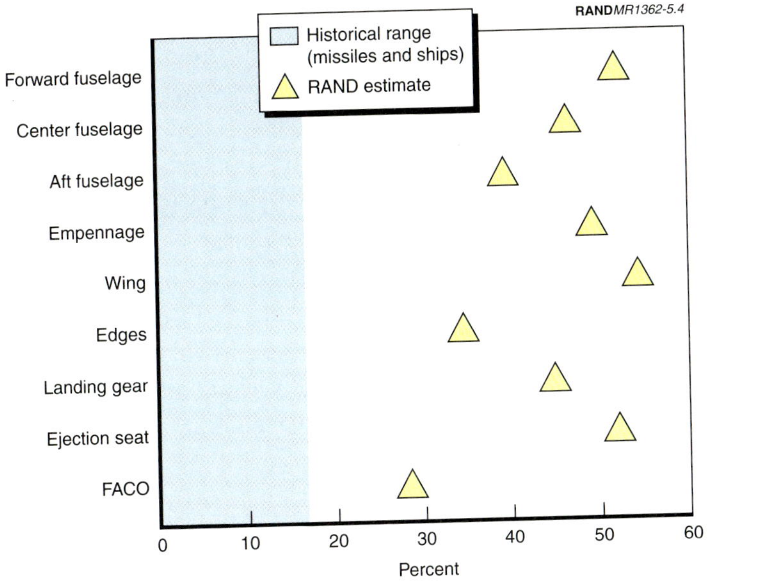

Figure 5.4—Airframe Component Break-even Estimates (Form-Fit-Function) from Table 5.5

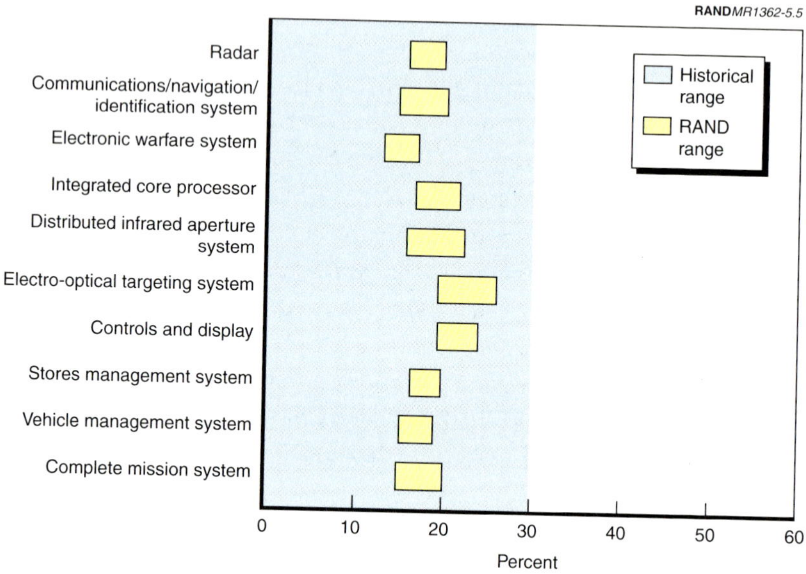

Figure 5.5—Mission System Break-even Estimates (Build-to-Print) from Table 5.6

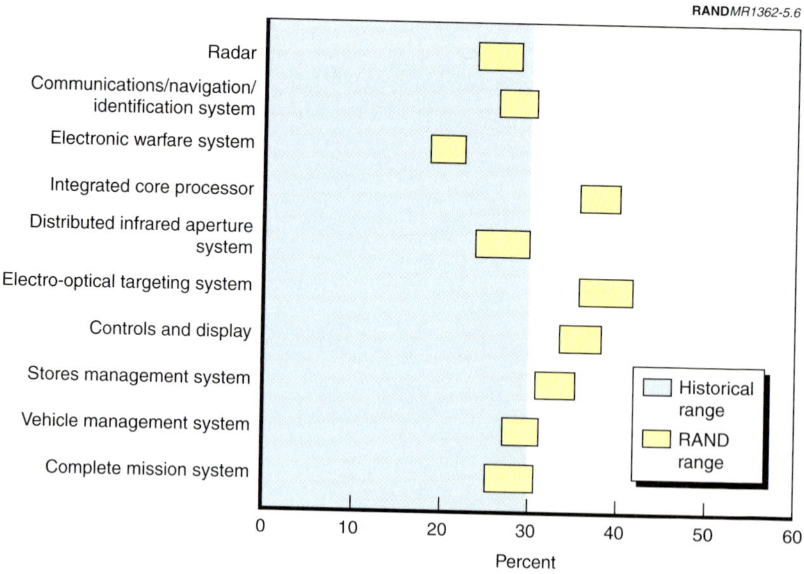

Figure 5.6—Mission System Break-even Estimates (Form-Fit-Function) from Table 5.6

duction effort. However, it is highly likely that the mission system equipment will have one or more major upgrades to keep pace with technology. If there is a major avionics upgrade after the first 1,000 aircraft are produced, then the extra investment costs and the loss of learning (for only 1,000 units) would have to be recovered over the first 1,000 units. This would roughly double the *RCR* needed to break even, thus significantly reducing the likelihood of being able to achieve adequate savings for the mission system equipment.

These comparisons of the savings *needed* to break even, with the savings *likely to be achieved*, lead to the conclusion that introducing competition to the production phase of the JSF program is unlikely to result in financial savings. If the decision on whether to introduce a competitive second source was to rely on that narrow criterion alone, the answer would almost certainly favor a single-source producer. However, it is conceivable that competition might lead to other benefits. In the next two chapters, we explore some of those possibilities; then, in Chapter Eight, we describe an integrated assessment of the several possible costs and benefits of near-term competition in the JSF program.

Chapter 6

Analyzing Operations and Support Costs

Competition during EMD and production can result in beneficial effects during operation and support of the system. History suggests that competition can lead to a better design and to increased quality control during production, both of which should lead to higher reliability during field operations. A frequently referenced example is the great engine war, which pitted General Electric's F-110 engine against Pratt & Whitney's F-100 engine to induce Pratt & Whitney to produce a more reliable version of the F-100 engine. Higher reliability should lead to some reduction in O&S costs. We therefore need to examine the extent of possible competition-induced reductions in O&S costs and see if such reductions might be large enough to affect our estimates of the likelihood of breaking even by introducing a second-source producer.[1]

Analytic Approach

Our analysis followed four steps:

- We identified the elements of O&S costs that are likely to be affected by contractor actions during EMD and production in a typical military aircraft program. This was done by reviewing the categories by which O&S costs are typically reported and judging which of those would be likely to change as a result of changes in system reliability.

[1] Competition in production might also lead to lower cost for replacement parts. However, analysis of lower spare parts costs requires a level of detail in O&S cost estimates that was not available at the time of this study.

- We determined the magnitudes of those competition-sensitive O&S costs in the JSF, as currently estimated for its projected operational life. The JSF Program Office provided those data.
- We calculated the sensitivity of those competition-sensitive O&S costs to changes in reliability. Those estimates, made by the Naval Air Systems Command (NAVAIR) using the JSF O&S cost estimation model, yielded a range of possible savings resulting from competition during production, expressed as a percentage change in certain JSF O&S costs.
- We used those savings to adjust the break-even calculations reported in Chapter Five to determine whether the projected O&S cost savings led to a significant change in the overall likelihood of breaking even.

Step 1: Identify Elements Most Affected by Contractor Action

Table 6.1 presents our assessment of the extent of contractor influence on the major categories of O&S cost in a typical fighter aircraft program, designated as high, medium, and low. The assessment reflects the judgment of personnel experienced in how the maintenance organization and process works. We judged only a few categories to be highly sensitive to changes in system reliability and, therefore, likely to be sensitive to the actions taken by the system producers.

We concluded that contractors have the highest level of potential influence over O&S costs in five areas—unit-level consumable supplies, depot-level reparables, airframe overhauls, engine overhauls, and support equipment repair. Our estimates of the percentages of the current O&S cost projections for the JSF that those elements represent are included in brackets in the table.

Before we go on to the second step, let us define what activities are included in each of these areas and their connection with reliability. *Unit-level consumption* consists of fuel and energy resources; operations, maintenance, and support materials consumed below the depot organizational level; reimbursement of stock fund for depot-level reparables (spare parts); operational munitions expended in training; transportation of materials, repair parts, and reparables between the supply or repair point and the unit; and other unit-level consumption costs, such as purchased services for equipment lease and service con-

Table 6.1
Degree of Contractor Influence over O&S Costs

O&S cost category	Contractor influence: High	Medium	Low	Percentage of JSF O&S cost
Mission personnel				23
Officers			●	
Enlisted		●		
Civilians		●		
Unit-level consumption				45
Petroleum, oil, and lubricants		●		
Consumable supplies	●			[9]
Depot-level reparables	●			[21]
Training munitions			●	
Intermediate maintenance				3
Officers			●	
Enlisted		●		
Civilian		●		
Contractor		●		
Consumable material/repair parts		●		
Depot maintenance				8
Aircraft overhaul				
Airframe	●			[5]
Engine	●			
Support equipment repair	●			
Sustaining support				10
Support equipment replacement		●		
Modifications		●		
Sustaining engineering services		●		
Software maintenance		●		
Simulator operations			●	
Technical publications			●	
Indirect support				12
Specialty training			●	
Permanent Change of Station			●	
Military medical care			●	
Installation support			●	

RANDMR1362-T6.1

NOTE: Numbers in brackets are included in the total for the main category.

tracts. *Consumable supplies* are those items purchased for one-time use, such as filters and oil, and then discarded when they must be replaced on an aircraft. *Depot-level reparables* are spare parts that, when removed from an aircraft, are tracked individually by an item number and returned to a central maintenance facility for repair and reuse. *Depot maintenance* is defined as personnel, material, overhead support, and depot-purchased maintenance required to perform a major aircraft or engine overhaul, and maintenance of a weapon system, its components, and support equipment at DoD centralized repair depots, contractor repair facilities, or on-site by depot teams.

Step 2: Determine the Magnitude of Competition-Sensitive O&S Costs in JSF

Many O&S costs depend on reliability. Reliability is significantly influenced by the contractor. Therefore, costs that depend on reliability are deemed to be highly influenced by the contractor. Aircraft that are required to be flown frequently, with little or no opportunity for maintenance between flights, should be designed to require fewer consumable supplies and repair parts. We did not include service maintenance personnel in the "high" category of contractor-influenced O&S costs, because maintenance manning at operational units is not strictly a function of aircraft repair frequency; rather, it is set by individual services on the basis of a host of factors. Theoretically, high-reliability weapon systems should reduce the need for maintenance personnel. Depot maintenance is also driven by the contractor-controlled airframe design and the need for periodic overhaul at centralized maintenance facilities. To the extent that the basic airframe design either obviates or reduces overhaul and major periodic maintenance, O&S costs can be reduced over those of historical systems. The design of an aircraft can reduce the amount of support equipment required to operate and maintain the aircraft, and higher reliability built into the support equipment itself can reduce the need to repair these items. Finally, the weapon system contractor, if given responsibility for centralized repair of the aircraft and support equipment, can reduce the cost of recurring repairs, even on highly reliable equipment.

The value for depot maintenance typically includes the engine. Because our study excluded the engine for which competition is already planned, we adjusted the value to reflect only the airframe, mission system, and support equipment repair. Table 6.2 shows the percentages of O&S costs for consumables, depot-level reparables, and depot overhaul

Table 6.2
Percentage of Airframe and Mission Equipment Operations and Support Costs Represented by Consumables, Depot-Level Reparables, and Overhaul

	Percentage of O&S costs	
	Airframe	Mission equipment
Consumables	~7%	~1%
Depot-level reparables	~6%	~7%
Overhaul	~5%	

RANDMR1362-T6.2

for just the airframe, mission system components, and support equipment repairs. The cost of depot overhaul for mission system and support equipment is small compared to the costs for the airframe and engine (in the range of 5 percent or less of the airframe estimated cost). Note that these values are for the total airframe and the complete mission system. Data were not available to do this analysis at the same level of detail as the production-cost *RCR* analysis in Chapter Five.

Step 3: Calculate the Sensitivity of Those Competition-Sensitive O&S Costs to Changes in Reliability

Figures 6.1 through 6.4 show the sensitivity to reliability changes of the largest cost elements from Table 6.1. These sensitivities were derived from data and JSF O&S cost model outputs provided by NAVAIR.

Step 4: Use Savings to Adjust Break-even Calculations of Chapter Five

Using the values from Figures 6.1 through 6.4, we determined the O&S dollar savings over the operational life of the JSF fleet, assuming a 30-year aircraft life. We then calculated the amount that the production-cost *RCR*s for the total airframe and the complete mission system package would be reduced if the reliability-driven O&S cost savings were realized. The results are shown in Table 6.3. The values in this table, as with all savings values shown previously, are undis-

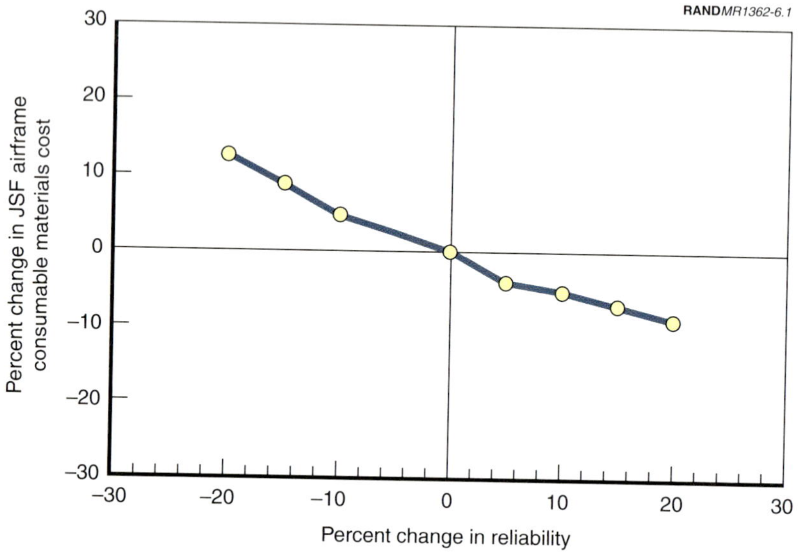

Figure 6.1—Sensitivity of Airframe Consumables to Reliability Changes

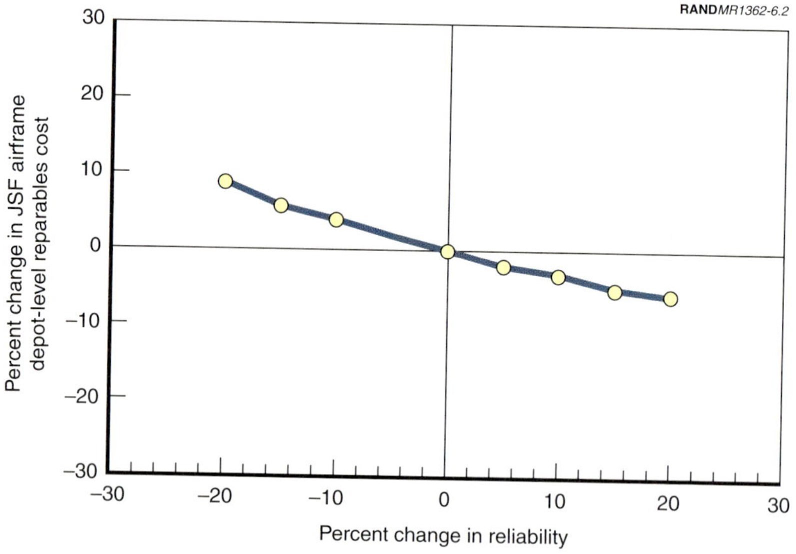

Figure 6.2—Sensitivity of Airframe Depot-Level Reparables to Reliability Changes

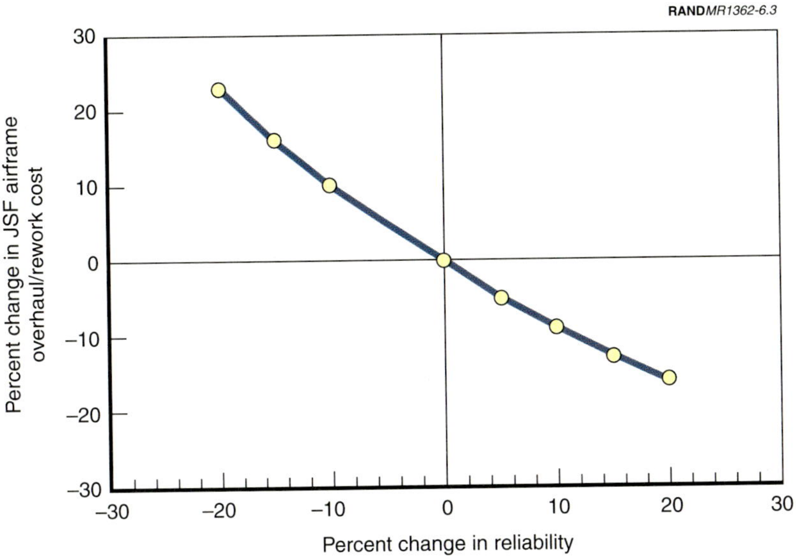

Figure 6.3—Sensitivity of Airframe Overhaul/Rework to Reliability Changes

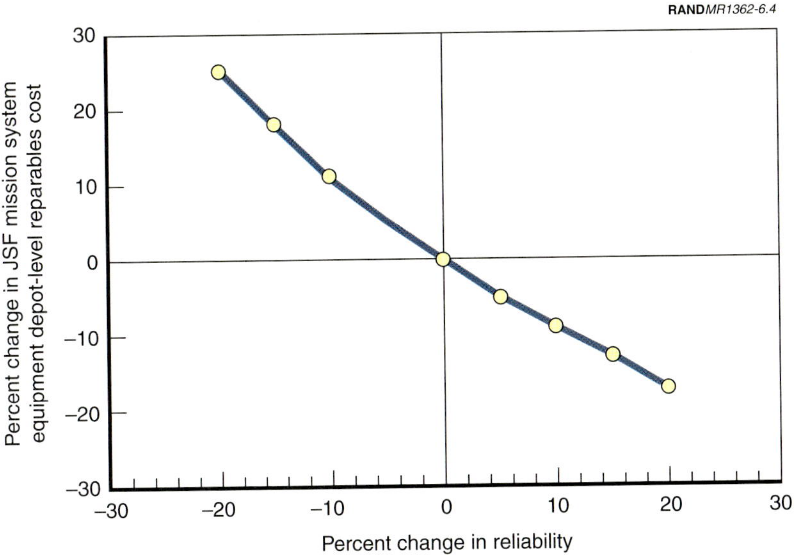

Figure 6.4—Sensitivity of Mission System Depot-Level Reparables to Reliability Changes

counted to enable comparison with the historical evidence. Because the O&S savings would be achieved several years further in the future than would savings in production cost, it is worth noting that corresponding discounted values are about 40 percent of the values shown in Table 6.3. For example, the –5.6-percent value for mission equipment with a 20-percent reliability improvement becomes –2.2 percent when discounted at 4.2 percent.

We conclude that competition-induced improvements in system reliability are likely to yield reductions in O&S costs, but that such reductions are not large enough to overcome the cost penalties of introducing competition.

Table 6.3
Effect of Reliability Improvements on JSF O&S Break-even Percentages

Reliability improvement (percent)	Delta from break-even baseline (percent)	
	Airframe	Mission equipment
5	–1.7	–1.6
10	–2.6	–2.9
15	–3.9	–4.3
20	–4.9	–5.6

RAND*MR1362-T6.3*

Chapter 7

Other Attributes of Competition

In the preceding two chapters, we examined the likelihood that the government would break even financially from introducing competition to the production phase of the JSF. There are several other widely acknowledged plausible effects of introducing competition in the JSF program—effects that are difficult to analyze in strict financial terms but that might be sufficiently important to justify their inclusion in an overall decision. Here we briefly examine five such effects:

- Would competition likely affect levels of program cost growth?
- Would competition strengthen the industrial base?
- Would competition reduce program risk?
- Would competition foster innovation and product quality?
- Would competition affect the level of international participation?

Would Competition Affect Likely Levels of Program Cost Growth?

We analyzed the record of development cost growth and procurement cost growth for several competitive and noncompetitive weapon system acquisition programs. Whether greater or lesser development cost growth indicates a more desirable outcome is arguable. Program success can be judged in terms of not just cost growth but procurement quantity changes. Unsuccessful programs will not be likely to have increases in quantity and may very likely have decreases. Successful programs will be the opposite, not likely to have decreases and more likely to have increases.

Historical Program Data

To assess the influence of competition on program cost growth and quantity changes, we drew on a RAND internal database of Major Defense Acquisition Program (MDAP) cost histories. For both development cost and production cost, we selected the estimate prepared at Milestone II (start of EMD) as the baseline and compared it with the actual costs incurred, adjusted for quantity changes. When determining actual costs, we examined only those programs that had at least some production experience (had passed Milestone III), and were at least five years beyond the milestone date.[1] We looked at programs that involved competition from the start of procurement, as well as programs for which competition was initiated after some amount of original prime-contractor production. Of the total 58 programs we examined, 14 involved competition and 44 did not. Single-Channel Ground and Airborne Radio System (SINCGARS) was the only electronics program involving competition. It was unique in that the second contractor's design was completely different from the initial design, an example of Form-Fit-Function. Table 7.1 summarizes the competitive programs. Table 7.2 summarizes the programs having no competition.

The baseline is significant because it relates to the quality of the estimate available and its implication for cost growth. At Milestone III, generally much more and better information is available than at Milestone II. Consequently, lower growth would be expected as measured from Milestone III than from Milestone II.

Procurement Cost Growth

The simple averages of the procurement cost-growth factors (CGFs) for the groups of systems in Tables 7.1 and 7.2 are shown in Table 7.3. One would expect competitive programs to have lower cost growth than noncompetitive programs, a hypothesis not borne out by the differences observed in Table 7.3. Furthermore, because of the small

---

[1] In some instances, the two milestones are coincident. Some ship programs begin with contracts for the lead ship (which is taken as EMD start), plus one or more follow ships (which is taken as the start of procurement). In other cases, there may have been no identifiable Milestone II cost estimate, so the cost track begins with Milestone III.

Table 7.1
Cost-Growth Data for Competitive Programs

	Baseline (FY01 $M)		Quantity		Cost-growth factors[a]	
Program	Development	Procurement	Baseline	Latest	Development	Procurement[b]
Electronics						
SINCGARS-V	**233.4**	**5,899.3**	**292,853**	**270,384**	**1.35**	**0.70**
Missiles						
AGM-65D	298.2	2,666.2	32,520	24,914	1.04	1.49
AIM-54C	182.1	389.8	465	2,483	1.67	2.42
AMRAAM	1,669.8	9,166.9	24,335	10,917	1.46	1.36
Dragon	294.5	1,720.2	246,575	67,561	1.88	2.29
Hellfire	616.5	821.1	24,600	45,659	1.33	1.77
Stinger—FIM 92A/92B	284.1	1,268.2	22,980	38,389	2.29	1.57
Tomahawk (B/R/UGM-109)	1,940.0	2,539.3	1,082	4,301	1.72	1.39
Ships						
CG 47	126.9	20,368.8	16	27	1.23	0.96
DDG 51	1,584.4	19,714.1	18	57	2.23	1.16
FFG 7	49.5	9,290.4	50	51	1.40	1.62
LHD 1	73.9	7,457.1	5	7	0.93	0.99
LSD 41 (Basic)	**93.4**	**5,451.3**	**12**	**8**	**1.08**	**0.88**
TAO 187	**23.9**	**3,809.9**	**17**	**16**	**0.97**	**1.06**

RANDMR1362-T7.1

[a]The cost-growth factors (CGFs) referenced to a Milestone III baseline cost estimate are indicated in bold to distinguish them from the Milestone II CGFs.

[b]The procurement cost-growth factors have been adjusted for quantity changes by adjusting the latest cost estimate from its corresponding quantity to the original quantity at the baseline estimate. The cost-improvement slope for the latest estimate is used.

sample sizes, none of the differences is statistically significant at the 10-percent level, as demonstrated by the ships and missiles CGFs considered both separately and together. When considered individually, the ships and missiles show higher average cost growth for competitive programs than for noncompetitive programs. When combined, the average competitive CGF is less than the noncompetitive value. A scatter diagram of the procurement CGFs, in Figure 7.1, indicates only one clear difference: Electronics programs have significantly lower cost and cost growth than do non-electronics programs.

Table 7.2
Cost-Growth Data for Noncompetitive Programs

	Baseline (FY01 $M)		Quantity		Cost-growth factors[a]	
Program	Development	Procurement	Original	Latest	Development	Procurement[b]
Electronics						
AFATDS	480.9	564.1	3,184	6,391	1.48	0.46
AN/TTC-39	414.0	1,582.0	292	100	1.54	0.96
CSSCS	136.0	153.6	1,031	3,081	1.45	0.48
JSTARS–CGS	348.5	663.3	95	121	2.25	0.95
Longbow Apache– FCR	916.0	596.0	227	320	0.98	1.80
MCS	**404.0**	**1,017.5**	**6,365**	**8,088**	**1.48**	**0.85**
SMART-T	238.3	682.4	364	320	1.32	0.71
Aircraft						
A-6E/F	**276.6**	**6,571.5**	**173**	**205**	**3.93**	**0.96**
B-1B	**4,375.2**	**30,771.1**	**100**	**100**	**1.31**	**0.98**
C-17	4,962.6	28,583.5	210	134	1.57	1.70
EF-111A	296.1	1,038.1	40	40	2.10	1.62
F/A-18 A–D	4,183.0	19,435.9	800	1,015	1.36	1.54
F-14A	3,883.9	19,858.1	463	583	1.53	1.25
F-15 A–D	7,078.6	18,431.3	729	1,074	1.48	1.47
F-16 A–D	1,683.4	11,270.6	650	2,201	2.51	1.29
JSTARS	1,865.5	1,576.8	10	14	2.20	2.04
T-45 Training System	742.7	4,182.1	300	234	1.53	1.74
Helicopters						
AH-64	2,266.3	4,803.9	536	811	1.20	1.75
CH-47D (Chinook)	221.4	2,392.9	361	474	1.13	1.36
Longbow Apache–AFM	396.9	1,178.4	227	530	1.93	2.19
OH-58D (AHIP)	347.7	2,330.0	578	382	1.13	1.85
UH-60A/UTTAS	1,396.2	6,381.0	1,107	1,327	1.16	1.36

RAND*MR1362-T7.2a*

Table 7.2—continued

	Baseline (FY01 $M)		Quantity		Cost-growth factors[a]	
Program	Development	Procurement	Original	Latest	Development	Procurement[b]
Missiles						
ACM	**2,117.9**	**5,782.9**	**1,461**	**460**	**1.08**	**1.41**
ALCM	1,749.2	5,734.6	3,424	1,763	1.38	1.03
ATACMS–Block I	884.8	631.2	1,000	2,299	0.97	1.35
Copperhead/ CLGP	318.0	2,189.9	132,650	24,545	1.23	2.11
Improved Hawk	**412.3**	**2,172.4**	**6,554**	**5,631**	**1.52**	**1.56**
Javelin (AAWS-M)	676.0	3,422.1	70,550	26,956	1.34	1.90
Longbow Hellfire	360.3	1,569.1	10,896	12,905	1.25	1.13
MMIII GRP	480.7	1,163.5	652	652	1.22	1.46
Pershing II	1,212.8	1,269.5	394	278	0.99	2.14
Roland	466.1	2,011.3	5,778	595	1.52	2.91
SADARM–155mm	314.9	1,071.8	64,123	50,000	1.53	1.95
TOW II	**161.7**	**3,226.7**	**141,224**	**142,429**	**2.85**	**1.15**
Ships						
MCM 1	**34.5**	**2,315.1**	**14**	**14**	**1.00**	**1.03**
MHC 51	**19.9**	**1,642.9**	**12**	**12**	**1.08**	**1.15**
Trident II (Sub)	**78.3**	**15,192.6**	**7**	**10**	**1.32**	**0.87**
Vehicles						
Bradley/IFV/ MICV	546.3	6,124.1	9,261	6,778	2.55	2.29
DIVAD/Sgt York	372.5	4,646.2	618	64	1.30	2.21
FAADS LOS-R (Avenger)	**15.1**	**1,400.2**	**1,207**	**773**	**3.26**	**1.13**
FMTV	74.1	8,160.3	118,935	86,916	1.68	2.39
M-1 (Abrams)	1,571.6	7,474.2	3,312	7,822	1.83	1.59
M-1A2 (Abrams)	**683.2**	**4,402.8**	**1,060**	**1,155**	**1.37**	**1.38**
MLRS/GSRS	**595.2**	**4,325.4**	**362,832**	**489,114**	**1.53**	**0.92**

RAND*MR1362-T7.2b*

[a]The cost-growth factors (CGFs) referenced to a Milestone III baseline cost estimate are indicated in bold to distinguish them from the Milestone II CGFs.

[b]The procurement cost-growth factors have been adjusted for quantity changes by adjusting the latest cost estimate from its corresponding quantity to the original quantity at the baseline estimate. The cost-improvement slope for the latest estimate is used.

Table 7.3
Simple Averages of Procurement Cost-Growth Factors

System	Competitive	Noncompetitive
Electronics	0.70	0.89
Ships	1.11	1.02
Missiles	1.76	1.68
Aircraft	—	1.46
Helicopters	—	1.70
Vehicles	—	1.70
Aircraft, helicopters, and vehicles	—	1.59
Ships and missiles	1.46	1.54
All non-electronic systems	1.46	1.57

RAND*MR1362-T7.3*

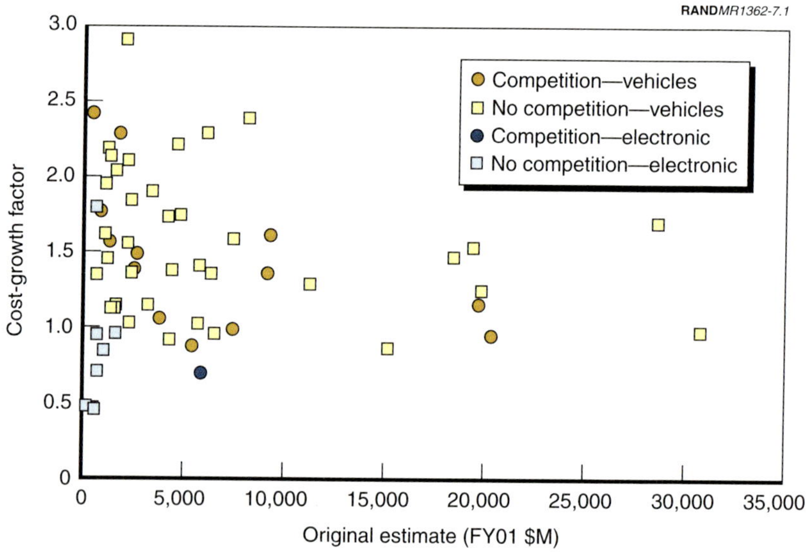

Figure 7.1—Scatter Diagram of Procurement Cost-Growth Factors

Our data contain no competitive aircraft, helicopter, or vehicle programs. However, note that for noncompetitive programs, the combined average procurement CGF for these systems, 1.59, is very close to the average for ships and missiles, 1.54—a very small difference.

Development Cost Growth

While procurement cost growth was of primary importance to this project, we also considered development cost growth. A scatter diagram of the development CGFs is shown in Figure 7.2, and the simple averages are presented in Table 7.4.

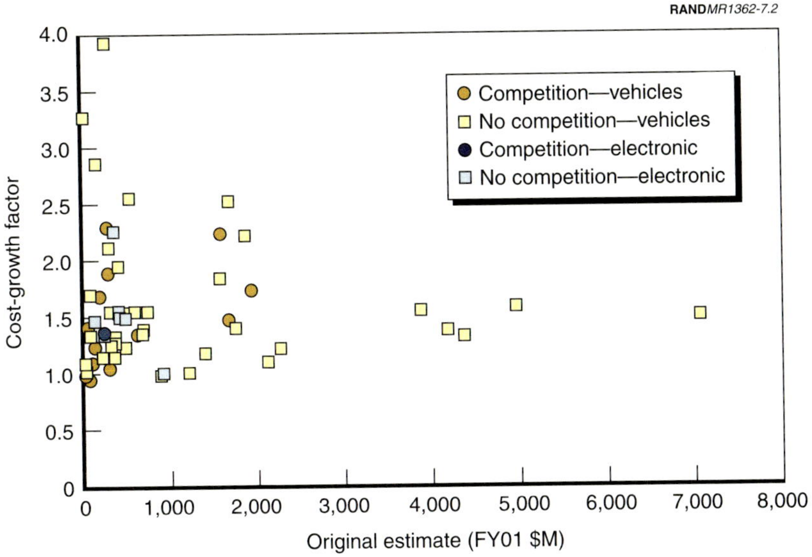

Figure 7.2—Scatter Diagram of Development Cost-Growth Factors

Table 7.4
Simple Averages of Development Cost-Growth Factors

System	Competitive	Noncompetitive
Electronics	1.35	1.50
Ships	1.31	1.13
Missiles	1.63	1.41
Aircraft	—	1.95
Helicopters	—	1.31
Vehicles	—	1.93
Aircraft, helicopters, and vehicles	—	1.80
Ships and missiles	1.48	1.35
All non-electronic systems	1.48	1.62

RANDMR1362-T7.4

Interpreting the differences between competitive and noncompetitive programs in terms of development CGFs is not as clear as for procurement CGFs. The data used to obtain the factors in this study are the total development cost for the program through all model changes of the weapon system. Also, most, if not all, of the programs included here did not involve competition during the development phase. Consequently, if competitive programs are more successful (i.e., have long production runs) than noncompetitive programs, we might expect development costs to grow, reflecting continued development of new models of the weapon system. This is countered by the hypothesis that successful programs will not have difficulties during development that lead to increased development costs. As with the procurement CGFs, the results are mixed and the differences between the competitive and noncompetitive development CGFs are not statistically significant at the 10-percent level.

Would Competition Strengthen the Industrial Base?

The defense industry has been undergoing consolidation and general reduction in capacity for more than two decades. Today, only three U.S. firms are capable of functioning as a prime in a major aircraft weapon system development and production program, down from nine in 1980. The number of first- and second-tier suppliers has likewise dwindled. The JSF is expected to be the only new major fighter aircraft development program in the next couple of decades. If all that business is granted to one firm, or even to a team consisting of two of the present three prime-level firms, there is concern that the loser or losers would be unable to sustain a competitive posture in the future, thus seriously reducing the breadth and depth of the U.S. fighter aircraft industrial base.

Prime Contractor

This issue was not addressed in depth in the present study. However, it has been examined in several other studies by RAND (Lorell and Levaux, 1998; Lorell, 1995; Drezner et al., 1992) and other organizations, although only a few of those resulted in formally published results. Results of those studies are mixed and generally not conclusive. There is broad agreement that vigorous competition by experienced firms contributes to the quality of the product. But exactly what constitutes a bare minimum in breadth of competition or whether the United States is nearing a critical point in the diminishing stock of combat aircraft developers is unclear. A brief survey of available information, published and unpublished, shows that no one has persuasively argued that a winner-take-all strategy in the JSF would inexorably lead to the loser exiting the fighter aircraft business, or even that the loser would clearly lose an important degree of capability to continue as a competitive supplier of combat aircraft systems in the future.

Note also that competition in *production* would have only second-order contributions to the industry base for *developing* future fighter aircraft. A production contract provides for sustaining engineering and enables expenditure of independent research and development funds for some technology development. However, it does not directly support the kind of staff or facilities needed for creative design of new systems. This issue is discussed more fully in Drezner et al. (1992).

We conclude that support of the industrial base at the prime-contractor level would provide an important, but not compelling, reason for sustaining competition during production of the JSF.

Subsystems

At the subsystem level, the issue is somewhat different, especially for certain kinds of electronic subsystems. In some cases, the industry is already down to a single supplier, and there are strong arguments in favor of encouraging another firm to enter those specialized areas. However, the need is for additional teams *developing* new technology and related system architectures. Again, this sort of work would not necessarily be supported by competitive *production* of the JSF. Indeed, the needed strengthening of the industrial base in these specialized areas would be better achieved in the near term by specifically supporting competitive development of subsystems that might be introduced into other combat systems and future variants of the JSF. This matter is discussed more fully in Chapter Nine.

Would Competition Reduce Program Risk?

Development of a new weapon system inevitably involves uncertainties and risks. The multiphase approach to such development, in which explicit attention is given to concept refinement and risk reduction before full-scale engineering development begins, helps to ameliorate risks but cannot eliminate them. Thus, any additional action or policy that could further control risks deserves consideration.

It is widely agreed that simply increasing the number of firms conducting parallel and independent development of a new system inherently reduces the overall risk of failure. If one firm has a 0.99 probability of success, then two equally qualified firms working independently on the same project should lead to an overall probability of project success of 0.9999 (assuming that the failure is uncorrelated). But that kind of benefit from multiple sources affects a project mostly in the development phase; competition during production appears to make little contribution toward overall project risk reduction.

Would Competition Foster Innovation and Product Quality?

As with project risk, the most effective way to foster innovation and improve product quality would be to introduce a second source during product development. A second source in production also might lead to improvement in product quality (through rigorous attention to quality control) and thus affect operational reliability of the product. Unfortunately, studies that explicitly examine the effects of second-sourcing on product quality are hard to find. Some such effect might be reasonably expected, but the extent, even the direction, of the effect is problematic. On the one hand, it can be argued that

- the second source would be starting the production learning process all over again and would suffer from quality problems that the original source had already solved.
- the increased emphasis on low production cost might create new quality problems for both suppliers.

On the other hand,

- the second source could benefit from most of the first source's experience and thus start out producing items of superior quality.
- enhanced competitive pressure would lead both suppliers to improve quality.

Measuring product quality can be contentious, because no single quantitative index of product quality is widely accepted. It would be desirable to examine several measures of quality for a variety of past programs to see if any of those measures could be correlated with the introduction of a second manufacturing source. In the time available for this study, we were able only to draw on earlier RAND research (Birkler et al., 1990). We could find no other relevant research. Birkler and his colleagues collected consistent and reliable data across several missile systems for only one index—flight test results—a limitation that is not as restrictive as it might seem. Flight test reliability information provides a good indication of whether a missile will hit a target, and so may be the best single parameter that can be used.

Tomahawk Experience

Here, we summarize the second-sourcing of Tomahawk production experience and the implications on flight test reliability. Tomahawk is the most complex second-sourced flight vehicle for which data are available. Given their greater complexity and longer flight times than for other missiles, cruise missiles offer the best available analog. The Advanced Cruise Missile (ACM) was also second-sourced, but the details of that procurement remain classified.

The primary reason for second-sourcing the Tomahawk was not to reduce cost but to improve system reliability. The Program Office was seriously concerned with Tomahawk reliability during system development and was not satisfied with the prime contractor's, General Dynamics Convair (GD/C) Division's, effort to deal with the problem. Since nothing gets management's attention as effectively as the prospect of having to share production with a competitor, the Program Office began thinking about second-sourcing early in the program.

It is difficult, however, to establish a cause-and-effect relationship between second-sourcing and improved quality. Quality improves over time in all programs as more tests are conducted and problems isolated. That was true in the Tomahawk program. The question is, Can a link be shown between second-sourcing and quality improvement?

We know that (1) uncorrected quality-assurance problems existed on and off between 1978 and 1982, (2) dual-source procurement was authorized in 1982, and (3) system reliability as measured by flight test results improved in the period 1983–1986. We also know that the Program Office did not rely exclusively on second-sourcing to influence GD/C to focus on the quality-assurance problem. The Defense Contract Administration Services Plant Representative Office (DCASPRO) issued five Method C corrective-action requests between November 1981 and May 1982 to deal with what the Program Office perceived as serious quality-assurance problems. In June 1982, DCASPRO issued a Method D corrective-action request, an action that is taken only "after sequentially exhausting every other avenue available by the Government to obtain corrective action by the manufacturer" (Birkler et al., 1990). Reliability showed a perceptible improvement by mid-1983, and overall missile reliability eventually achieved a level comparable to that of other missile programs.

Figure 7.3 displays Tomahawk (GD/C) long-term flight test reliability[2] through almost 180 flights, at which point it had achieved a reliability level of about 97 percent. Unlike the other missile systems examined, its flight test reliability was not maintained at the 90-percent level until the 75th flight. The other missiles attained that level much earlier in their flight test histories—TV Maverick at test 6 and IIR Maverick at test 18.

Tomahawk flights 27 through 78 have an average reliability of 87 percent. Flights 80 through 177 have an average reliability of about 96 percent. This difference is statistically significant at the 95-percent confidence level. There is some reason to believe that GD/C's effort to

[2] Birkler et al., 1990. The model used for the estimation and forecasting of reliability was the Lloyd-Lipow model, which calculates overall system reliability as a function of the number of failures and the number of tests. Failures are weighted to accommodate the small sample sizes and the presumed capability to improve the design in the remainder of the development phase, thus not repeating a failure mode.

RANDMR1362-7.3

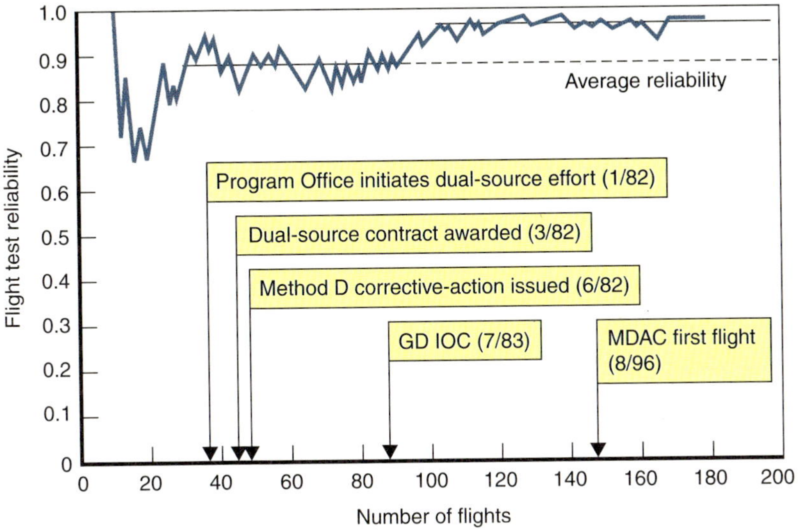

Figure 7.3—Tomahawk Missile Flight Test Reliability Trend

improve reliability began to show results at about the time of flight test 75 in March 1983. Events before that date that could have affected GD/C's attitude toward quality assurance are as follows:

- January 1982: Cruise Missile Program Office initiates dual-source effort.
- November 1981–May 1982: Five Method C corrective-action requests issued.
- March 1982: Dual-source contract awarded.
- June 1982: Method D corrective-action issued.

According to the Program Office and contractor personnel, both second-sourcing and the corrective-action requests were important. However, the proximate cause for action appears to have been the DCASPRO letter of June 24, 1982, implying drastic action if "the Government's missile program is not measurably improved" in the next 120 days. During that grace period, GD/C formed a corporate team to identify and correct quality-assurance problems, and three of the five

Tomahawk flights between November 1982 and March 1983 were successful (one of the two failures was apparently related to the McDonnell Douglas (MDAC) land-attack guidance system). Reliability improved steadily thereafter. The problem all along, according to the Program Office, was GD/C's disinclination to take the strenuous action required to correct engineering and manufacturing quality-control problems. Imminent competition and the more direct corrective-action requests both contributed to a change in corporate attitude. It would be mere speculation to claim that one was more important than the other, however.

Does second-sourcing result in a more reliable product? According to the Tomahawk Program Office, improved quality assurance, not cost reduction, was the primary reason for second-sourcing in the Tomahawk program. There is little doubt that competition forces an attitude change on contractors who may have grown complacent in single-source production. When reliability is an important source-selection criterion, it has to be taken seriously.

Since our evidence comes from a single program, we cannot say definitely that equipment produced in second-source programs has a superior reliability record.

Would Competition Affect the Level of International Participation?

Maintaining competition during the production phase could clearly affect foreign industrial participation in the JSF program. Traditionally, acquisition programs with significant foreign participation have included carefully negotiated formal government-to-government agreements delineating the participation of specific foreign firms and allocating work share and work tasks. In past international collaborative programs, the value of the work allocated to foreign industry usually closely approximated the financial contribution of the participating governments. Changes in any of the industrial aspects of a program typically would require a major renegotiation of the government-to-government agreement. Thus, in a traditional international collaboration program, introducing competition in the production phase could raise serious issues regarding possible changes in work share, work allocation, and work tasks, as agreed in the original government-to-government MoUs.

JSF has broken out of the traditional mold of collaborative programs. For the EMD phase, JSF planners have agreed that there will be no guaranteed foreign work share based on government-to-government agreements. No detailed discussion of specific industry participation, work share, or work tasks will be included in government-to-government agreements. All foreign participants recognize that there are no guarantees that work share will be proportional to the participating government's contribution to EMD.

Foreign industry participation is determined and managed entirely on the industry level. Foreign industry participation is available through the prime contractors, subcontractors, and vendors. Foreign companies that wish to participate in the program compete with U.S. companies, as well as with companies from third countries, for work share and work tasks. Companies are selected on a best-value basis, whether they are foreign or domestic. Strong incentives for this approach are provided by the ongoing competition between the two prime contractors and the intense pressure to meet very demanding URF cost goals established by the Program Office from the earliest phases of the program.

Nonetheless, foreign participating governments have a recognized expectation of receiving an equitable return on their investment through industry participation in the program. Both of the competing prime contractors recognize the importance of including significant industrial participation on their teams representing the countries that are participants in the Concept Demonstration Phase and that are likely to play a major role in EMD and production (see Tables 7.5 and 7.6). Both prime contractors recognize that they would be politically at a competitive disadvantage in the downselect if their teams did not include equitable industrial representation from the key foreign government participants in the program—particularly the United Kingdom, since the UK formally has been granted a position of influence over the final downselect.

Not surprisingly, both the Boeing and the Lockheed Martin teams include significant representation from British firms. Interestingly, the proportional value of the program that will go to British firms is likely

Table 7.5
Companies Teamed with Boeing During CDP

Company	Role
Aerosystems	Logistics support
Alcoa	Castings
BAE Systems	Fuel system, cockpit displays, electronic warfare
B.F. Goodrich	Fuel system
Cytec Fiberite	Composites
Dowty	Actuators
EDO	Weapon bay swing arm system
FHL	Attitude control
Flight Refuelling	Fuel system
Fokker	Wire bundles, aerostructures
GE Aircraft Engine	Alternate engine
Harris	Network cards
Hexcel	Composites
Honeywell	Subsystems
IBM/Dassault	Information systems
Martin Baker	Ejection seat
Moog	Actuators
Parker	Fuel system
Raytheon	Mission systems
Rolls-Royce	Vertical lift systems, alternate engine (F120 with GE)
Smiths Industries	Stores management system
TRW	Navigation, communications
United Technologies	F119 engine, subsystems

RAND*MR1362-T7.5*

SOURCE: Merrill Lynch & Co., Global Securities Research & Economics Group, Global Fundamental Equity Research Department, "The Pilot-No. 13: The Joint Strike Fighter Program," *A Monthly Global Overview of the Aerospace & Defense Sector,* 6 October 2000, p. 23.

NOTE: The suppliers and products shown are believed to be accurate as of October 2000. However, these arrangements are subject to change as the program evolves. The list is therefore intended to be representative, not definitive.

to be higher than the proportional value of the British government contribution. In other words, British industry is actually likely to get a better deal than it would have in a traditional collaboration program, for which industrial participation is exactly proportional to the foreign

Table 7.6
Companies Teamed with Lockheed Martin During CDP

Company	Role
BAE Systems	Aft fuselage, flight controls, vehicle management computer, electronic warfare
B.F. Goodrich	Landing gear
Cytec Fiberite	Composites
EDO	Weapons stores
Fokker	Wiring
General Electric	Alternate engine
Harris	Electronics
Hexcel	Composites
Honeywell	Power management
Kaiser Electronics	Displays
Litton	Electronics warfare
Martin Baker	Ejection seat
Moog	Actuators
Northrop Grumman	Center fuselage, electro-optical sensors, radar
Rolls-Royce	Lift fan, ducts, alternate engine (F120 with GE)
Smiths Industries	Stores management system, electrical power management system, vehicle management subsystems
TRW	Communications, navigation, actuation control
United Technologies	F119 engine, subsystems

RAND*MR1362-T7.6*

SOURCE: Merrill Lynch & Co., Global Securities Research & Economics Group, Global Fundamental Equity Research Department, "The Pilot—No. 13: The Joint Strike Fighter Program," *A Monthly Global Overview of the Aerospace & Defense Sector*, 6 October 2000, p. 20.

NOTE: The suppliers and products shown are believed to be accurate as of October 2000. However, these arrangements are subject to change as the program evolves. The list is therefore intended to be representative, not definitive.

government contribution. For example, on the Lockheed team, BAE Systems is treated as a "Partner," along with Northrop Grumman; on the Boeing side, many British firms are part of the Boeing "One Team" unified management structure.

However, the specific British companies involved and the work tasks they undertake vary significantly between the two prime contractor teams—the result of the prime contractors and other suppliers having been granted the authority of determining foreign industry participation on their own, based on their own criteria and best business practices. Foreign firms were selected on the basis of best-value standards. Thus, the specific companies and specific work tasks vary between the two teams, but the overall foreign representation by country is roughly the same. There is also equitable participation on both teams by firms from all the other countries participating in the CDP. This participation is likely to continue during EMD and into production.

Therefore, the revolutionary way that the JSF international program provides incentives makes it highly unlikely that recompetitions during the production phase will dramatically affect overall industrial participation by foreign companies: In any new competition, the primes will again recognize that equitable representation of foreign participants' firms on their teams will be a factor in deciding which team wins the competition. Neither team will want to risk losing the competition by being perceived as ignoring equitable representation of key foreign participating governments' firms. However, the enormous competitive pressures to meet demanding URF cost goals will ensure that the primes and their suppliers will select foreign firms according to best-value principles and technical needs, not merely on politics.

In conclusion, although competition during the production phase could lead to a change in specific foreign firms and specific foreign work tasks, competition of any type will still preserve an equitable industrial share in the program for the major participating foreign governments. Indeed, it appears that the incentives in the program could even lead to overrepresentation of foreign firms, unlike in the traditional approach of officially mandating that industrial participation remain precisely proportional to foreign government contribution to the program. Therefore, we conclude that competition during the production phase poses no serious problems for international participation.

Chapter 8

Near-Term Policy Options for Competition in the JSF Program

We have examined, with varying degrees of rigor and completeness, the major program features that might be affected by introducing competition in the production phase of the JSF. We now must assemble those disparate pieces of information and draw an overall conclusion on how such competition might be introduced, and on the balance of advantages and disadvantages resulting from such competition.

Competitive Production of the JSF

The immediate and direct policy question is, Should action be taken during development of the JSF to establish a competitive production source for part, or all, of the JSF weapon system? We assume such competitive production would start with the LRIP units and extend throughout the JSF production program.

We have analyzed the expected effects of such competition on production costs, O&S costs, and future cost growth. We have also examined several other possible consequences of competition, and we have examined those issues for competition in production of the airframe and the mission equipment. In some areas, the results appear conclusive; in other areas, they appear less so. Those individual results and interpretations must be integrated into an overall assessment.

Any such integration must be subjective. The answer will depend on the relative weighting given to the various consequences of competition. A useful method for conducting a subjective integration is the stoplight chart, in which each element is assigned one of the four colors. The results of our analyses are summarized in Figure 8.1 using this display strategy. Our interpretations are described below.

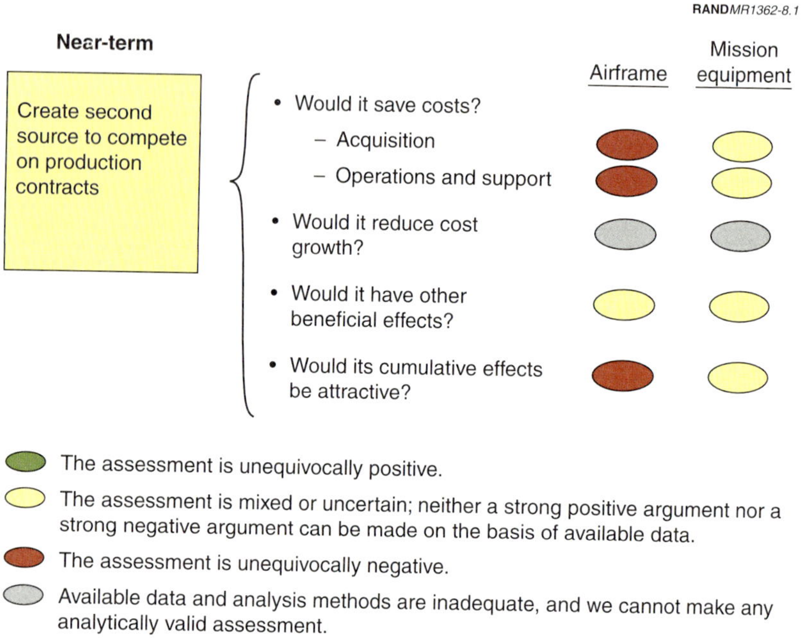

Figure 8.1—Summary Assessment of Effects from Introducing Competition

Would It Save Acquisition Costs?

Airframe: RED. There is only about one chance in five of recovering the investment necessary to establish a competitive producer for airframe and major subsystems.

Mission Equipment: YELLOW. Our analysis shows that there is a better than an even chance of recovering the investment needed to establish a second supplier of mission equipment if the second source builds-to-print following the design of the winning prime. However, achievement of significant savings is far from assured. If the second source was directed to design its own components to a Form-Fit-Function specification, the chance of recovering investment slips to only slightly better than even, and that strategy would complicate logistics support of the system. However, those estimates were made on the assumption that the design would be reasonably stable over the full 3,002-unit production run. If the mission equipment will be upgraded periodically, then the cost of introducing competition would have to be recovered over smaller production runs, thus increasing the *RCR* needed to break even. Because of the inherent uncertainty regarding such future actions, no strong argument can be made for or against competition on the basis of acquisition cost savings.

Would It Save Operations and Support Costs?

Airframe: RED. Competition might lead to improvements in system reliability, thus reducing O&S cost, but not nearly enough to project a good chance of recovering investment.

Mission Equipment: YELLOW. Reliability improvements have a stronger effect on reducing O&S costs, but not enough to recover investment. Combining such potential O&S savings with the analysis of production cost improves the overall chance of recovering investment, but still leaves considerable uncertainty. Again, no strong argument can be made for or against competition on the basis of O&S cost savings.

Would It Reduce Cost Growth?

Airframe and Mission Equipment: GRAY. The historical data are so sparse and mixed that no reliable conclusion can be drawn.

Would It Have Other Beneficial Effects?

Airframe and Mission Equipment: YELLOW. Competition would probably yield some beneficial effects through strengthening the industrial base and moderating overall program risk, and there is anecdotal evidence of improvements in innovation and product quality. However, the quantitative historical evidence for those benefits is weak or non-existent. We find no reason to expect that competition would have important effects on the level of foreign participation in the JSF program. We judge the overall result as tending toward positive, but still being basically inconclusive.

Would Its Cumulative Effects Be Attractive?

Airframe: RED. There is a strong probability that competition would not lead to savings large enough to pay back the front-end investment. Those potential net costs are not outweighed by the other potential benefits of competition for which the expected consequences are uncertain or neutral.

Mission Equipment: YELLOW. This assessment is less definitive than for the airframe. Might the combination of several "neutral" elements lead to a positive judgment? We think not, for the reasons spelled out below.

If the competition were Build-to-Print, the chance of overall cost savings through reduction in production costs and O&S costs would be better than even, maybe even approaching two chances out of three—but still not a strong expectation of significant net cost reductions. Further, any benefits to industrial base, risk reduction, and innovation likely would be relatively small because the second source would simply be replicating the winning team's design. There is no strong expectation of overall benefits.

If the competitor developed its own design to a Form-Fit-Function specification, the only practical approach would be to develop an entire mission equipment set rather than one or more individual elements, such as a radar (the rationale for this judgment is discussed in Chapter Nine). Such a development's additional cost means that it would have only slightly better than an even chance of breaking even. The potential benefits to the industrial base, risk reduction, and innovation are stronger than for the BTP option but still not enough to build a conclusive argument supporting the additional investment of several billion dollars.

The probability of achieving net savings is further diminished by the expectation that the mission equipment will be upgraded during the life of the JSF, thus reducing the production quantity over which the investment would have to be recovered.

Overall Conclusion

Our overall, net conclusion is that there is no persuasive basis at this time for investing in a second source for production of all, or any part, of the JSF weapon system. This conclusion does not apply to the engine, which was not included in our analysis.

Chapter 9

Other Options for Competition

In Chapter Four, we outlined an overall analysis strategy encompassing four categories of competition that might be applied to the JSF program. The figure depicting those categories is repeated here as Figure 9.1. The majority of the analysis, reported in Chapters Five through Eight, focused on options for near-term competition within the JSF program. In this chapter, we discuss options in the remaining three areas depicted in Figure 9.1.

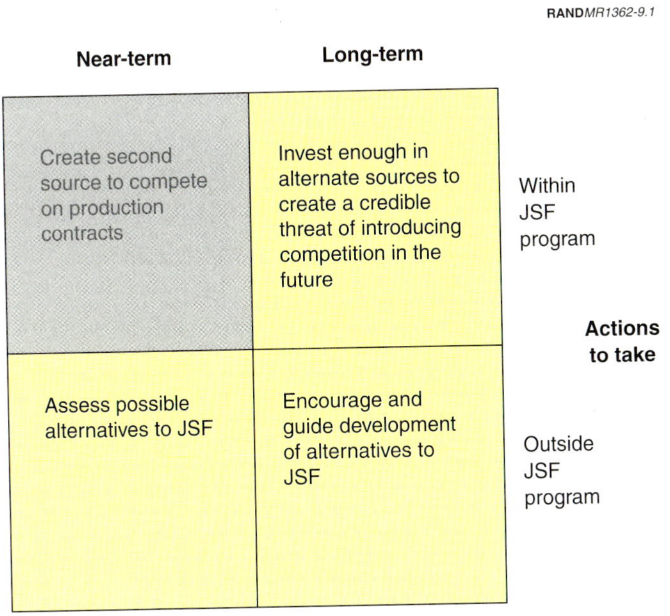

Figure 9.1—Other Near-Term and Long-Term Competition Options Within and Outside the JSF Program

Options for Long-Term Competition Within the JSF Program[1]

The production phase of the JSF is expected to extend over a period of roughly 25 years, during which some 3,000 aircraft will be delivered to the U.S. services. If history is a guide, we can expect several new variants to be introduced during that time as mission needs evolve and new technologies become available. The timing and exact content of those upgrade programs cannot be defined at this time. However, it seems likely that they will offer opportunities for competition in both design and production activities. It seems useful to look ahead and project as best we can the nature of such future competitions and to identify how they might affect the JSF program and any near-term actions needed to enhance the effectiveness of those options in the future.

In principle, any competition option examined in the preceding chapter could be applied in the future. However, history suggests that aircraft vehicle configurations do not evolve radically throughout the life of a fighter aircraft design. Such airplanes typically grow in gross weight, and modest refinements are made in aerodynamic performance and in details of structure design, but the basic vehicle design remains relatively static. Furthermore, the analysis of the vehicle structure and major subsystems described in Chapters Five and Six showed little promise of breaking even financially through introduction of a second production source, and there is no apparent reason to expect that fact to change in the foreseeable future. We therefore see little opportunity to introduce competition in the future for production of vehicle structure and major subsystems.

Mission equipment presents more opportunities for future competition. The most important reason for this expectation is that mission system components are likely to undergo relatively rapid evolution in both mission needs and enabling technologies. By the time the JSF enters operational service, roughly ten years from now, there is a strong probability that the operational concept will have changed somewhat, redistributing emphasis on the various elements of the electronics suite of sensors, displays, and associated computation methods. Further-

[1] Our analysis of the JSF mission equipment described in this section drew extensively on an ongoing U.S. Air Force–sponsored RAND Project Air Force study of JSF avionics cost-estimating relationships and cost-reduction initiatives.

more, the relevant technologies are advancing rapidly, thus enabling improvements in capabilities over the suite now being developed for the initial variants of JSF.

For these reasons, we expect that new designs for mission equipment will emerge over the next decade. Does it make sense to introduce a second source that could compete for design and production of that new mission equipment suite? Consideration of a second source is justified because the potential improvements in JSF mission effectiveness through upgrades in mission equipment are so large that they justify the stimulus to technical innovation that is believed to be achieved through competition. *The argument does not have to rely entirely on potential cost savings; some increase in cost might be justified as investment in improved mission capability through technological innovation.*

To devise an overall concept for competition in design and development of future mission equipment, we need to examine three issues:

1. Can individual elements of the mission equipment suite be competed or must such competition encompass an entire avionics suite?
2. Should the second source compete on the basis of BTP or FFF?
3. What should be done in the near term to enable future competition?

We examine each in turn, drawing on features of the mission equipment designs being offered in the present JSF competition.

Integrated Design Versus Separate Subsystem Design[2]

Mission system avionics will perform a variety of functions in the JSF. They will

- control the plane
- monitor its resources
- provide information that pilots need to perform their missions and to control and/or guide weapons to targets
- provide communication links to report mission status.

---

[2] In our analysis, we examined individual elements of the mission system in detail. Due to proprietary constraints, that analysis cannot be fully reported here.

In the past, contractors working on conventional aircraft programs typically have developed discrete requirements for each avionics subsystem. But with the JSF, each prime has analyzed the JSF mission operational requirements and has developed an integrated mission equipment package in which traditional subsystems take on nontraditional functions. Driven both by the JSF's complex mission requirements and by desires to save costs, contractors are using combinations of subsystems to accomplish tasks heretofore performed by single subsystems. For example, traditionally, radar functions have been performed by a single stand-alone subsystem. Contractors have parceled out some of the JSF's radar functions among the integrated core processor (ICP), electronic warfare (EW) components, and other subsystems. Moreover, the EW and the communications/navigation/identification friend or foe (CNI) subsystems share antennas, an integrated core processor—which replaces most of the separate processors in each subsystem—and common support electronics.

Each of the two prime contractors working on the JSF expects that such integration of mission system avionics will enable it to significantly reduce these components' cost and weight and boost their performance. However, the two designs differ in how functions are distributed across different hardware and software elements and in how they are integrated into the flight vehicle. The differences vary from minor to major, directly affecting the ability of a subsystem vendor to enter into a future competition against the firms participating in the initial production program. Given the design strategy being employed, traditional methods of fostering avionics competition may not be effective.

Reverting to a conventional federated, or stand-alone, subsystem design would facilitate future competition at a component level. However, with such a conventional design, the overall program cost would be greater than with an integrated mission system package. This does not mean some level of traditional competition for each avionics subsystem is futile. What it means is that a detailed assessment of each subsystem is required before the decision to compete the subsystem (or a part of it) can be made. The decision is not whether another vendor could make its subsystem fit in the winning team's design. Rather, a hypothetical decision to fit a competitor's component into the existing JSF design would need to be assessed on the following factors:

- How much would it cost to make the new design fit?
- What innovations would the proposed new design bring to the existing mission system?
- What else would be needed to foster competition for the entire avionics package?

Our preliminary analysis of these issues indicates that the Defense Department should consider a new approach to fostering competition for the mission system at a future upgrade: Compete the entire mission system package.

The integrated systems approach enables significant cost, size, and weight savings for the mission system package. Competing at the mission system level would empower the competitor to be innovative and develop systems that enhance mission performance while minimizing cost. In addition, it would help preserve the avionics integrator industrial base.

Should Competition Be in FFF or BTP Mode?

In Chapter Five, where we examined near-term opportunities for competition, we considered options in which the second source would competitively build a component or subsystem using the original source's design (Build-to-Print) or design and manufacture its own variant that would fit into the existing overall system (Form-Fit-Function) as a direct substitute for the original source's design. Which approach should be followed by our suggested second source for future mission equipment?

In principle, it should be possible for the prime to design a follow-on mission equipment suite, then a second source could compete for production on a minimum-cost basis. However, we argued earlier that a major future benefit from a mission equipment second source would be to encourage innovation and exploit advances in technology. The advantage being sought here is different from that of the near-term competition modes examined in Chapter Five; here, we are primarily seeking design improvements, not just reductions in manufacturing cost. To achieve such benefits, it would be necessary for the second source to create its own design for all or parts of the mission equipment suite.

Such an FFF strategy would require close cooperation between the overall system prime and the competitive second source for mission equipment, because the second-source design would have to fit into the existing flight-vehicle and overall-weapon-system design.

What Near-Term Actions Should Be Taken to Support Future Competition for Mission Equipment?

One strategy the DoD could employ would be to wait a few years until changing mission needs and evolving technologies justify an upgrade in mission equipment, then evaluate the option of introducing a competitive second source. However, that might not provide the maximum benefits. Near-term stimulus might not be available to support appropriate technology innovation at the desired rate, and some key firms might have left the industry. Thus, we want to consider sustaining a second source at a level less than full EMD but sufficient to ensure technology advances and retain a vigorous industrial base to support future competition.

How might such a sustainment phase be structured? We suggest some guidelines. Following the arguments outlined above favoring an integrated design approach, it would be appropriate to organize a "shadow mission system design team" to develop a next-generation mission system. A *shadow team* would have as its strategic goal to create a strong competition to the JSF prime contractor team for design and production of the second-generation mission equipment suite. The strength of that competition would depend on how well the shadow team used available resources to develop critical technologies, nurture the firms best capable of providing such technologies, and integrate the results into a next-generation design concept.

How much might such a shadow team cost? We can draw on some related research in which RAND researchers examined the minimum size at which an aircraft design team must be sustained through lean times in order to be competitive on the next design competition. That study (Drezner et al., 1992) identified five categories making up the overall activity needed to sustain a vigorous design capability:

- Technology development
- Engineering and management staff
- Facilities
- Financial support
- Institutional structures and management organizations.

RAND's survey of several aircraft-development firms over several cycles of lean and full business activity suggested that "the historical minimum size that design organizations reached (was) characterized by about $100 million in annual funding and 1,000 people in engineering and technical management."

We have not conducted a similar survey of modern military avionics development firms, but we have no reason to believe that a team preparing to compete for development of a future mission equipment suite for the JSF would be larger than an equivalent aircraft design team. Thus, we estimate that the annual sustainment cost for such a team would be roughly $50 to $100 million, depending on the range of system elements subject to vigorous development activities.[3] This investment rate would equal some 10–20 percent of the annual cost of full-scale EMD for the mission system.

Assuming that such a shadow competitor might be sustained for five years before deciding whether to implement a full competition for an upgraded mission system (and hence the need to invest in a full EMD for the competitor) would yield a total investment of maybe $250 to $500 million, or roughly 10 to 20 percent of full EMD cost. But that would not necessarily represent a net outlay. It seems reasonable to assume that the suppliers of mission system equipment to the JSF program would view such a program as a threat of future competition, and they might consequently work to reduce their own prices and, hence, be better prepared to meet such future competition.

How large would such price reductions have to be in order to cover the shadow competitor costs? Making a limit-case assumption that the price reduction would affect the entire production run of mission system equipment, we obtained the results shown in Table 9.1. Of course,

[3] Some believe that, because of the need to keep 6–10 individual contractors active, our cost estimates are too low.

Table 9.1
Production Savings Needed to Break Even on Investment in Shadow Competitor

Investment in shadow competitor (percent of full EMD)	Prime cost reduction needed to break even (percent)
10	2
20	4

RAND*MR1362-T9.1*

if the price reduction affected only a later segment of the production, the required reductions needed to break even would be greater. But even if the recouped cost was less than the amount necessary to fully cover the investment in the shadow competitor, the benefits of having a full design competition for the next upgrade appear to justify the investment.

The government would get four benefits from such an annual investment:

- It would be assured of continued development in technologies and system architectures directly relevant to the JSF. Having such a program in place would provide the institutional structure necessary to support and guide appropriate development work.
- It would be able to selectively support a few key firms that might otherwise leave the business of developing fighter aircraft mission equipment, and thus strengthen the industrial base for future programs.
- The shadow competitor would serve as a backup in case the prime developer of mission equipment "stumbled" in some important way.
- The existence of such a program might provide to the prime developer of mission equipment a sense of impending future competition in this important area, thus improving the quality of the main JSF development program.

The complete implementation of such a strategy has not been fully examined in the present study. This option deserves careful consideration, but its size, scope, and structure will depend on more-detailed examination.

Options for Competition from Outside the JSF Program

There are many near-term and longer-term options for competition from outside the JSF program. However, a very quick survey of these options suggests that none is particularly attractive, for a variety of reasons.

Near-Term Options

In the near term, modifications or upgrades of some existing aircraft could in some ways be viewed as competitive with the CTOL variant of the JSF. On the one hand, a multirole air-to-ground variant of the F-22 Raptor, now completing EMD, could certainly equal or surpass the JSF CTOL in most performance areas. On the other hand, a variant of the large, two-engine F-22 would undoubtedly be considerably more expensive than the JSF CTOL, assuming the JSF program achieves its URF cost goals. An upgraded variant of either the F-16D Block 60+ or the F-15C or F-15E, including JSF avionics subsystems, also might equal or exceed the JSF CTOL in many performance areas. Indeed, an advanced F-16 is already being developed for the United Arab Emirates and includes the latest AESA radar technology and other advanced avionics, as well as conformal fuel tanks added to the fuselage for greater range. However, both F-16 and F-15 upgrades would likely fall well short of the JSF CTOL in low-observability (LO) characteristics, as well as in other areas. Furthermore, an F-15 variant would probably surpass the stringent JSF URF cost goals.

For carrier use, an upgraded variant of the F/A-18E/F now going into production is an obvious alternative to the JSF CV. Indeed, the Navy is already adding an advanced AESA radar and other improved avionics to the existing F/A-18. However, this option suffers from the same problems as the CTOL alternatives: The two-engine F/A-18 variant would likely be more expensive and would lack the LO characteristics of the JSF CV variant.

There is no plausible near-term alternative to the JSF STOVL variant. The only possibility would be a major developmental program to upgrade the existing subsonic AV-8B Harrier, an aircraft developed in the late 1970s. Of course, such a variant would also probably lack the LO characteristics of the JSF STOVL design.

Perhaps most important, these options would mean the loss of the substantial cost-improvement curve benefits of producing large numbers of variants of the same basic design that is at the heart of the cost savings approach of the JSF program. Instead of all three services procuring essentially the same basic airframe design and the same mission avionics, three separate modification and upgrade programs of different existing aircraft would have to be undertaken, each resulting in the procurement of far smaller numbers of unique designs. Thus, the economies of scale sought in the JSF program would be lost.

Variants or upgrades of foreign fighters currently under development or in production could also be considered possible alternatives to the JSF, although probably only for foreign customers. Possible alternatives for the JSF CTOL variant include the UK-German-Italian-Spanish Eurofighter, which is now just entering into production; the French Dassault Rafale, also just entering production; and the Swedish Saab Gripen. The Eurofighter and the Rafale, high-performance modern fighters, show some of the same disadvantages of the F-15 and the F/A-18: They are both expensive two-engine aircraft with poor LO characteristics relative to the JSF designs. The Rafale was designed from its inception for both land-based and carrier options. The first-production Rafale is the Rafale M variant for the French Navy, which could be considered as an alternative for the JSF CV. However, the problems mentioned above for U.S. fighters apply to the Rafale M. Already in production, the Gripen is a smaller, lighter, less capable F-20–class fighter whose design does not possess the basic performance and LO characteristics of the JSF designs. No STOVL candidates exist in Europe, other than a major upgrade of the existing subsonic Harrier GR.7. Finally, none of these aircraft would provide the benefits of the huge economies of scale sought in the JSF program by using a single basic design to meet all services' needs.

Long-Term Options

In the longer term, more radical modifications of the already-mentioned existing fighters could be considered as alternatives to the JSF. However, the scale of modifications necessary to be truly competitive with the JSF would amount to developing essentially a new aircraft. Thus, no particular advantage would necessarily be gained from taking this route.

In the long term, radical alternatives such as Unmanned Combat Air Vehicles (UCAVs) might be substituted for some missions envisioned for JSF. Indeed, the U.S. Defense Department recently awarded a contract to Boeing to continue development of its UCAV design. Other radical concepts such as mounting larger numbers of smart standoff munitions on aerial missile launch platforms such as existing or future bombers, or even modified commercial wide-body airliners, might appear attractive. The type of munitions that could be used might be something like an upgraded Joint Air-to-Surface Standoff Missile (JASSM) with an active target-recognition sensor for greater accuracy against high-value targets, even when launched hundreds of miles from the target area. The JASSM is currently in full-scale development for the Air Force.

An even more radical concept might be to place weapon-launching platforms in space and launch precision-guided munitions from there against land and sea targets. This, as well as many other similar suggestions, is obviously an option that might become available only far in the future.

The problem with most of these radical options is that the ultimate effectiveness and capabilities of such future systems are not known at this time. The UCAV and JASSM are being developed, but real capabilities and effectiveness of even the first-generation systems that are currently under development are not known. No well-developed doctrine or operational concepts exist for the substitution of air-breathing or space-based smart-munition launch platforms for the traditional combat missions assigned to fighter-attack aircraft such as the JSF. Furthermore, the ultimate cost competitiveness of such systems with the JSF cannot be calculated at this time.

Conclusion

We conclude, therefore, that if the JSF program achieves its system performance and URF cost goals, no realistic competitor exists, at least in the short term. Existing stealthy aircraft that might be competitive are too expensive, whereas existing nonstealthy aircraft will lack the operational capabilities of JSF. However, if JSF URF costs begin to escalate, other platforms might become feasible alternatives, especially for the CTOL JSF variant. The most obvious would be a downgraded ground-attack variant of the F-22 or an upgraded F-16 Block 60+. An enhanced variant of the F/A-18E/F would probably be the most likely candidate for the CV of the JSF.

If JSF costs begin to escalate, careful and extensive combat modeling must be conducted to determine the cross-over point between cost and effectiveness, at which candidates such as an upgraded F-16 or a downgraded F-22 become viable substitutes. At this stage of the JSF program, competition from outside the program appears to be limited.

Chapter 10

Conclusions and Recommendations

The JSF's original plan calls for the DoD to select, at the beginning of Engineering and Manufacturing Development, a single winner to develop and produce this next-generation fighter aircraft. DoD officials are concerned about the suitability of that plan and have asked RAND to examine alternatives.

Our objective in this study was to define a range of opportunities and options open to DoD to introduce competition into the JSF program and to assess the balance of advantages and disadvantages offered by each option. The options included near-term actions that might be taken in 2000–2001 and longer-term actions that might be implemented later in this decade or even beyond, while the JSF is still in production. Except for the engine, which is currently being competed, all aspects of the JSF program were included in the study.

Several criteria are commonly applied to such an analysis:

- Reducing production cost
- Reducing operations and support costs
- Reducing cost growth throughout the program
- Encouraging innovation and quality
- Strengthening the industrial base.

The quantitative analyses we performed were primarily focused on the first two criteria; the other criteria were considered, but at a lesser level of detail.

We examined each competition option, some in more detail and more quantitatively than others, constrained only by the available time (about four months) and the effort level enabled by our client, the Under Secretary of Defense for Acquisition, Technology and Logistics. Throughout the study, we obtained critical information from the competing program contractors, and we obtained additional data from the Joint Strike Fighter Program Office.

From our analyses, we make two recommendations:

- *Stick with the winner-take-all strategy for near-term EMD and production of the JSF.*

Despite the potential advantages that might accrue, we estimate that establishing a competitive production line for part or all of the JSF weapon system would require a front-end investment, together with increases in recurring costs, that probably would not be recovered through price reductions that might result from competitive forces. Even when considering all possible benefits from such competition, we are unable to make a persuasive case for establishing a competitive production source.

As a corollary, the absence of competitive pressures makes it vital that all government agencies involved in managing the JSF program use every available strategy to control program costs and to ensure high quality standards during production. A wide range of contractual and regulatory procedures is available to help achieve these goals.

- *Policymakers should examine the costs and benefits of supporting a second industry team so that it could be capable of competing to develop and manufacture the next major upgrade of the mission system equipment.*

The mission system constitutes the eyes and ears of the JSF and provides a powerful contribution to its overall combat effectiveness. Many of the enabling technologies are evolving much more rapidly than are most flight-vehicle technologies, and we can expect several major upgrades in mission system during the life of the JSF. It therefore makes sense to begin planning now for the first upgrade.

We suggest investigating the establishment of a "shadow" industry team to begin developing system architectures and component technologies that would be tailored to the JSF platform but would be focused on technological advancement, cost reduction, and any new mission requirements that might emerge. The team would be ready to provide competition for EMD on an upgraded mission system at some time in the future. This strategy would ensure that future managers have the option of a competitive second source for designing, and possibly producing, future mission equipment upgrades, an option deemed valuable in the present stage of JSF development but that might not otherwise be available in the future.

Whether it will appear attractive to fund EMD and production by a second source in the future cannot be predicted at this time. There are too many uncertainties: the amount of progress made by the shadow team, performance of the system prime to that point, general economic conditions, etc. Our rough estimates suggest that such a shadow team could be supported in the near term at 10–20 percent of the annual cost of a full-scale EMD program for the mission system. Such an investment is not trivial, but it appears to warrant consideration.

Our charter was to seek ways to inject competition into the JSF program. At the present time, a shadow team appears to be the most attractive option. Within the scope of our study, we were unable to examine the idea in detail. Therefore, we recommend only that the notion be examined with an appropriate level of care and detail by the JSF management.

Bibliography

Archibald, K. A., et al., *Factors Affecting the Use of Competition in Weapon System Acquisition*, Santa Monica, Calif.: RAND, R-2706-DRE, February 1981.

Asher, Harold, *Cost-Quantity Relationships in the Airframe Industry*, Santa Monica, Calif.: RAND, R-291, July 1, 1956.

Beltramo, Michael N., *Dual Production Sources in the Procurement of Weapon Systems: A Policy Analysis*, Santa Monica, Calif.: RAND, P-6911, November 1983.

Birkler, J. L., et al., *Issues Associated with Second-Source Procurement Decisions*, Santa Monica, Calif.: RAND, R-3996-RC, December 1990.

Birkler, J. L., and J. P. Large, *Dual-Source Procurement in the Tomahawk Program*, Santa Monica, Calif.: RAND, R-3867-DR&E, 1990.

Daly, G. G., et al., *The Effect of Price Competition on Weapon System Acquisition Costs*, Arlington, Va.: Institute for Defense Analyses, P-1435, September 1979.

Dews, Edmund, Giles Smith, et al., *Acquisition Policy Effectiveness: Department of Defense Experience in the 1970s*, Santa Monica, Calif.: RAND, R-2516-DRE, 1979.

Drezner, Jeffrey, et al., *Maintaining Future Military Aircraft Design Capability*, Santa Monica, Calif.: RAND, R-4199-AF, 1992.

Drinnon, J. W., and J. R. Hiller, *Predicting the Costs and Benefits of Competitive Production Sources*, The Analytical Services Corporation, Report 1511, December 1979.

Flynn, Brian, and Dennis Herrin, *Analysis of Competitive Procurement of Selected Navy Weapon Systems*, Second Edition, NCCA, 1989.

Greer, Willis R., Jr., and Shu S. Liao, *Cost Analysis for Dual Source Weapon Procurement*, Monterey, Calif.: Naval Postgraduate School, NDS54-83-011, October 1993.

Kratz, L. A., et al., *Establishing Competitive Production Sources: A Handbook for Program Managers*, Fort Belvoir, Va.: Defense Systems Management College Press, August 1984.

Lorell, Mark A., *Bomber R&D Since 1945: The Role of Experience*, Santa Monica, Calif.: RAND, MR-670-AF, 1995.

Lorell, Mark A., and Hugh P. Levaux, *The Cutting Edge: A Half Century of U.S. Fighter Aircraft R&D*, Santa Monica, Calif.: RAND, MR-939-AF, 1998.

Lovett, E. T., and M. G. Norton, *Determining and Forecasting Savings from Competing Previously Sole Source/Noncompetitive Contracts*, Army Procurement Research Office, APRO 709-3, October 1978.

Margolis, M. A., R. G. Bonesteele, and J. L. Wilson, "A Method for Analyzing Competitive, Dual Source Production Programs," presented at the 19th Annual DoD Cost Analysis Symposium, September 1985.

Pilling, Donald L., *Competition in Defense Procurement*, Washington, D.C.: The Brookings Institution, 1989.

Rich, Michael D., *Competition in the Acquisition of Major Weapon Systems: Legislative Perspectives*, Santa Monica, Calif.: RAND, R-2058-PR, 1976.

Schmall, Joseph H., *Introduction to Defense Acquisition Management*, Fort Belvoir, Va.: Defense Systems Management College Press, June 1996.

Tyson, Karen W., et al., *Acquiring Major Systems: Cost and Schedule Trends and Acquisition Initiative Effectiveness*, Arlington, Va.: Institute for Defense Analyses, P-2201, March 1989.

—, *The Effects of Management Initiatives on the Costs and Schedules of Defense Acquisition Programs*, Arlington, Va.: Institute for Defense Analyses, P-2722, November 1992.

Washington, William N., "A Review of the Literature: Competition Versus Sole-Source Procurements," *Acquisition Review Quarterly*, Spring 1997, pp. 173–188.

Zusman, M., et al., *A Quantitative Examination of Cost Quantity Relationships, Competition During Reprocurement, and Military Versus Commercial Prices for Three Types of Vehicles*, Arlington, Va.: Institute for Defense Analyses, S-429, March 1974.